MATHS ALIVE!

Also available from Cassell:

L. Burton, *Girls into Maths Can Go*
L. Burton, *Gender and Mathematics*
A. Orton, *Learning Mathematics*, 2nd edition
A. Orton and G. Wain, *Issues in Mathematics Education*
D. Thyer, *Mathematical Enrichment Exercises*

Maths Alive!

INSET Mathematics for the National Curriculum (Key Stages 1, 2, 3)

Edith Biggs and Kathleen Shaw

CASSELL

Cassell
Wellington House
125 Strand
London WC2R 0BB

215 Park Avenue South
New York
NY 10003

British Library Cataloguing-in-Publication Data
A catalogue record for this book is available from the British Library.

ISBN: 0-304-32994-0 (hardback)
0-304-32990-8 (paperback)

Typeset by York House Typographic Ltd.
Printed and bound in Great Britain by Redwood Books, Trowbridge, Wiltshire

Contents

Introduction

This book has been written for all the teachers in primary, middle and lower secondary schools who want to become familiar with and understand mathematics at different Key Stages of the National Curriculum for September 1995. It should be of great value to Mathematics Co-ordinators in primary schools who are responsible for organizing INSET sessions. This book will not only help them to plan their school's INSET sessions but also provide teachers with reinforcement, since they can refer within this book to the problems and activities relevant to the children they teach.

In addition, the book will be very useful to students at Colleges of Education and at Departments of Education in Universities when planning classroom activities, to their lecturers and to advisory staff of local education authorities (inspectors, advisers and advisory teachers) who will be interested in the development of mathematical concepts, and, in fact, to all those concerned with helping children between the ages of 3 and 13 years to learn and to enjoy mathematics.

This book is particularly concerned with progression in mathematics throughout the primary and lower secondary years, i.e. Key Stage 1 (Levels 1 to 3), Key Stage 2 (Levels 2 to 5), and Key Stage 3, which covers Levels 5 and 6 of the National Curriculum. Pupils who show much interest in mathematics and are generally high attainers may cover the Programmes of Study at Level 5 and even at Level 6 while they are still at the primary school. We have therefore included activities at Levels 5 and 6 to ensure that these very able pupils are stretched in the subject and enabled to achieve their maximum potential. On the other hand, some pupils at the top of the primary school may not be able to complete all the activities for Level 4.

Most teachers responsible for covering these stages are not specialists in mathematics; some may not have enjoyed the subject themselves at school. Moreover, the National Curriculum in mathematics includes a number of topics – algebra, handling data (statistics and probability), geometry (shape and space) and the use of calculators and computers – not previously taught to this age group. At the same time, teachers are expected to ensure that their pupils have a more secure and extensive number knowledge.

The inclusion of new topics has caused many teachers to be apprehensive of recent developments in the National Curriculum. While the reduction in January 1991 of the number of Attainment Targets (from 14 to 5) has eased the teachers' task of assessing their pupils' progress at all levels, the resulting changes in the Statements of Attainment may well have confused many. For this reason, mathematical topics have been grouped whenever possible (to emphasize progression within a concept), and relevant Statements of Attainment listed. These, and the Programmes of Study, are sometimes expressed in unfamiliar language, but the examples suggested, particularly in the most recent order for September 1992, serve to clarify the descriptions.

In December 1991, the Department of Education and Science published *Strands in Attainment Targets for Mathematics*. Because every strand at all the levels for the five Attainment Targets is concisely delineated on a single poster-sized sheet of paper, teachers can see at a glance what is expected of their pupils. This poster is a useful framework for the Programmes of Study and Attainment Targets in mathematics.

In the text, statements from the National Curriculum are often displayed and italicized.

INSET

Because of the changes in mathematics which the National Curriculum necessitates at all levels, the provision of INSET assumes greater importance then ever before. However, very few local education authorities (LEAs) have been able to maintain their advisory services in mathematics at their former favourable strength. At present, most INSET sessions take place within individual schools or are shared between small groups of schools. School-based INSET does have many advantages: it gives teachers the opportunity of working together with their colleagues, on activities specified by the National Curriculum which are unfamiliar to them.

INSET sessions should be planned to cover a number of aspects.

1. *Working through the activities* prepares teachers for any problems their own pupils might meet.
2. *Planning sequences of activities* ensures that there is progression throughout, for children of a variety of abilities. At one extreme, there are children who have difficulty in acquiring new concepts and need more time to work through more, and more varied, mathematical activities. At the other extreme, there are pupils who are quick to learn mathematics, and who are highly motivated. These pupils must not be held back; they should be given more challenging problems based on the work in hand, as well as being moved on to programmes of a higher level.
3. *Devising questions* to help pupils make progress and *preparing assessments* to enable teachers to find out whether individual pupils have acquired the concept under consideration are tasks best undertaken on a group basis, before the teaching starts.
4. *Acquiring the necessary mathematical background*, in order to understand the mathematical objectives, is essential for each individual teacher.

Further follow-up INSET sessions are also needed for teachers to report back on progress made in their own classrooms, to discuss problems which have arisen, and to plan future activities.

The format of this book

First and foremost, this book is concerned with INSET in mathematics and its application in classrooms.

We have included problems which teachers could use to introduce each new topic or concept, in order to stimulate children to work out a solution for themselves. A series of INSET activities at different levels is provided to help teachers to solve the problems when the topic is also new to them. They should compare and assess different solutions.

Adaptations of the INSET activities are provided to suit classrooms at different levels. We include questions which teachers can ask their pupils to help them to learn, and questions which will help teachers to assess whether or not the pupils have acquired the underlying concept.

New vocabulary is included, since one of the assessments should be designed to determine whether the pupils use essential vocabulary correctly, and wherever the mathematical background is likely to be unfamiliar to teachers, it is included, sometimes as an introduction, at other times in conclusion. All the activities have been tried out in primary school classrooms and modified as necessary.

Mathematics Co-ordinators

Mathematics Co-ordinators need continued support in their efforts to foster the development of mathematics in their schools. On appointment, many of them will have been released part-time from their classrooms to take a course in Mathematics Education at primary/lower secondary levels; these courses were governed by the principles set out clearly in the Non-statutory Guidance. On returning to their schools, the co-ordinators needed time to inform themselves: of the state of mathematics teaching throughout their own school; of strengths and weaknesses; of teachers who would welcome help in their classrooms when introducing a new topic; and of those who needed to extend their teaching methods but were reluctant to do so, and would therefore need sympathetic handling. Indeed, one of the most difficult tasks facing the Mathematics Co-ordinator is to help teachers change from a prescriptive style of teaching (in which methods are taught and practised until they are memorized) to a more open style (in which pupils discuss problems, exchange ideas and undertake joint investigations). The many suggestions included in this book should serve to assist the Mathematics Co-ordinator to succeed in providing the best INSET for their school's needs, and raise the standards of mathematics teaching at this level.

THE NATIONAL CURRICULUM: THE CHANGES

The National Curriculum was introduced by the Government in its Education Reform Act (1988) in order to raise the standards of children's education between the ages of 5 and 16 years. In July 1987 a working group was appointed to give advice on appropriate Attainment Targets and Programmes of Study for three core subjects – English, mathematics and science – and seven foundation subjects, of which some are options. Teachers in state-maintained schools are required to teach the subjects in the National Curriculum, to assess children's progress and to inform parents of their progress.

In June 1989 it was proposed that there should be fourteen Attainment Targets and fourteen Programmes of Study for each subject. These were written in parallel. In mathematics, some practical examples were also included for clarification. The introduction of each Attainment Target began with the children who entered the first year of compulsory schooling in 1989–90. The other Attainment Targets were to be phased in gradually.

Early in 1991 it was decided that the original proposals for the National Curriculum were urgently in need of review, since teachers were finding it difficult to cover the Attainment Targets and the necessary assessments. The fourteen Attainment Targets were therefore reduced to five. This reduction was achieved by a simplification of the targets rather than by a reduction in content. Despite this simplification, teachers' associations recommended that their members should boycott the tests because they made such heavy additional demands on teachers.

In March 1993, Sir Ron Dearing was asked by the Secretary of State for Education and Science and the Secretary for Wales to undertake a review of the framework of all subjects in the National Curriculum. It was felt that the basic aims of the National Curriculum and the assessment arrangements were being undermined by its excessive content, complexity and over-prescription. Sir Ron Dearing was asked to simplify the National Curriculum and to give maximum support for classroom teachers. Once again the content has not undergone a major change, but Attainment Targets and Programmes of Study have been clarified, and the number of Targets has been further reduced.

In Key Stage 1 (Levels 1, 2 and 3) there are three Targets: Using and Applying Mathematics; Number; Shape, Space and Measures. In Key Stage 2 (Levels 2, 3 and 4) there are four Attainment Targets: Using and Applying Mathematics; Number; Shape, Space and Measures; Handling Data. In Key Stage 3 (Levels 3, 4 and 5) they are: Using and Applying Mathematics; Number and Algebra; Shape, Space and Measures; Handling Data. It has also been suggested to teachers that Programmes of Study, rather than Attainment Targets, should guide the day-to-day planning of their teaching and their assessment of pupils' progress. Moreover, the Levels have been extended for each Key Stage as follows: Key Stage 1 – Levels 1 to 3; Key Stage 2 – Levels 2 to 5; Key Stages 3 and 4 – Levels 3 to 10. It is hoped that the overlap of levels will encourage pupils who have ability in mathematics to reach their full potential.

Throughout this book we have referred to Programmes of Study rather than to Attainment Targets. The Targets have been replaced by ten levels for each subject, which are usefully listed at the end of the book. We have carefully compared the proposals for the Mathematics Curriculum published in 1991 with the draft published in May 1994. Where we considered that the 1991 contribution contained useful items not specified in 1994 we have included them in this book. One other change was the removal of mathematical content which the committee felt did not underpin mathematics at higher levels. There were two items, networks and three-dimensional graphs; we have now omitted these.

During the review, teachers were asked to send comments to the chairman on the proposed changes; 1400 schools and teachers did so. Final modifications will be made in September 1995, after which date it is strongly recommended that no further changes should be made for another five years.

Chapter 1

Using and Applying Mathematics: Programmes of Study

This Programme of Study is concerned with the process of learning mathematics by using and applying the subject. It covers the methodology for all the Programmes of Study and sets the scene for the varied processes of learning mathematics in all its aspects. This chapter therefore makes a most important contribution to the proper learning of mathematics.

What do we mean by mathematics?

Mathematics is a network of ideas. Within this network of strands, all the Programmes of Study find a place and cohere to make for progress in mathematics. Teachers and pupils together explore the network, finding links between familiar and newly acquired mathematics. The teachers' task is to organize and provide experiences which enable pupils to make these links for themselves and thus to develop their understanding.

Mathematics can also be thought of as a search for patterns and relationships (in number, algebra and shape) and the representation of the findings in words and pictures, ordered tables and graphs. Mathematics is therefore a subject in which pupils must be encouraged to take an active part in learning and to think for themselves.

METHODS OF TEACHING

The National Curriculum documents encourage teachers to help pupils to develop their own methods of calculation. Formerly, most teachers showed pupils one preferred method (usually decided by the school) and set them to practise it until it became an automatic procedure. Problems often developed later, when part of the method had been forgotten; then, with little understanding of what they were asked to do, pupils floundered and could not reconstruct the method.

Case 1.1 Many 10- and 11-year-olds forget how to carry out division of fractions because they were taught to 'invert and multiply' without really understanding why the procedure worked. If they had first understood that division and multiplication are inverse operations – in other words multiplication undoes division and division undoes multiplication – they might have understood why the formula worked. Consider the example:

$3\frac{1}{2} \div \frac{1}{4}$

Here the problem can be expressed as: 'How many quarters are there in $3\frac{1}{2}$?' Inverting the divisor and multiplying we have:

$3\frac{1}{2} \times \frac{4}{1}$,

which gives 14 as the answer to the problem.

Some teachers find it difficult to make the transition from prescribed methods to a variety of methods from which pupils choose the one most appropriate to each situation. Throughout this book some suggestions are provided, specifically to help these teachers.

Progression from one level to the next is clear-cut, as are the different methods of representation: oral and pictorial, by ordered tables and by graphs.

We now concentrate on Key Stage 1 of this section. The initial statement is:

> The sections of the programmes of study interrelate. Developing mathematical language, selecting and using materials, and developing reasoning should be set in the context of the other areas of mathematics. Sorting, classifying, making comparisons and searching for patterns should apply to work on number, shape and space, and handling data. The use of number should permeate work on measures and handling data.

This general statement is followed by a detailed description.

1. Pupils should be given opportunities to:
   a. use and apply mathematics in practical tasks, in real-life problems and within mathematics itself;
   b. explain their thinking to support the development of their reasoning.

Pupils should be taught to:

2. Make and monitor decisions to solve problems;
   a. select and use the appropriate mathematics;
   b. select and use mathematical equipment and materials;
   c. develop different mathematical approaches and look for ways to overcome difficulties;
   d. organise and check their work.
3. Developing mathematical language and communication:
   a. understand the language of number, properties of shape and comparatives, e.g. 'bigger than', 'next to', 'before';
   b. relate numerals and other mathematical symbols, e.g. '+', '=', to a range of situations;
   c. discuss their work, responding to and asking mathematical questions;
   d. use a variety of forms of mathematical presentation.
4. Developing mathematical reasoning:
   a. recognise simple patterns and relationships and make related predictions about them;

b. ask questions including 'What would happen if?' and 'Why', e.g. considering the behaviour of a programmable toy;
c. understand general statements, e.g. 'all even numbers divide by 2', and investigate whether particular cases match them.

For many children this will be the end of their first two years of schooling; next year they will begin junior school with a new teacher. Let us take a look at Key Stage 2 for Using and Applying Mathematics, to note the rate of progression.

Key Stage 2

The general statements begin in the same way:

> The sections of the programmes of study interrelate. Developing mathematical language, reasoning and skills in applying mathematics should be set in the context of the other areas of mathematics. Measurement should be associated with handling data and shape and space. Calculating skills should be developed in number and through work on measures and handling data. Algebraic ideas of pattern and relationships should be developed in all areas of mathematics.

You will notice that there is a good deal of common ground in the two general statements. It looks as though it should not be too difficult to plan Key Stage 2 so that it develops smoothly from the achievements of Key Stage 1. Contact between the teachers at both stages is important to organize. Let us now look at the details of Key Stage 2.

1 Pupils should be given opportunities to:
   a. use and apply mathematics in practical tasks, in real-life problems and within mathematics itself;
   b. take increasing responsibility for organising and extending tasks;
   c. devise and refine their own ways of recording;
   d. ask questions and follow alternative suggestions to support the development of reasoning.

2. Making and monitoring decisions to solve problems:
   Pupils should be taught to:
   a. select and use the appropriate mathematics and materials;
   b. try different mathematical approaches; identify and obtain information needed to carry out their work;
   c. develop their own mathematical strategies and look for ways to overcome difficulties;
   d. check their results and consider whether they are reasonable.

3. Developing mathematical language and forms of communication:
   a. understand and use the language of number; the properties and movement of shapes; measures; simple probability; relationships, including 'multiple of', 'factor of' and 'symmetrical to';
   b. use diagrams, graphs and simple algebraic symbols;
   c. present information and results clearly, and explain the reasons for their choice of presentation.

4. Developing mathematical reasoning:
   a. understand and investigate general statements, e.g. 'wrist size is half neck size', 'there are four prime numbers less than 10';

b. search for pattern in their results;
c. make general statements of their own, based on evidence they have produced;
d. explain their reasoning.

Once again teachers will be aware of the close similarities between the detailed requirements described at Key Stages 1 and 2. The headings for the different sections are almost identical. This should help teachers in their planning for many children at the end of the junior school. Able children should certainly progress to work from Key Stage 3 and a few to work from Key Stage 4. At these stages, once again there are close resemblances in the headings at Key Stage 2 and Key Stages 3 and 4. In the latter stages the emphasis is on

1. problems that pose a challenge;
2. considering different lines of mathematical argument;
3. finding ways of overcoming difficulties;
4. developing and using their own strategies;
5. breaking complex problems into a series of tasks;
6. using mathematical forms of communication, including diagrams, tables, graphs and computer print-outs;
7. interpreting mathematics presented in a variety of forms; evaluating solutions;
8. examining critically, improving and justifying their choice of mathematical presentation.

INSET

It is important that teachers should experience, with their colleagues, the processes of solving real-life problems and carrying out investigations. Only then will they have enough confidence to try this method when teaching their pupils. They also need to become familiar with the different ways of representing their findings.

MATERIALS NEEDED FOR THE CLASSROOM

To ensure successful progress towards Programme of Study 1, it is essential to provide a wide variety of materials, e.g. empty cartons of various sizes, shapes and materials, pieces of fabrics of different kinds and patterns, and newspapers and magazines. Subsequently, in the classroom, if children are encouraged to join in the collection and organization of these materials, they are more likely to understand their potential, as well as becoming familiar with their whereabouts.

The school will also need to provide:

- paper of different colours and sizes;
- paint of different colours with brushes of various types, and crayons;
- constructional materials as well as interlocking cubes;
- scissors, paste and glue;
- hundreds, tens and units material;
- top-pan balance scales.

The emphasis given throughout Key Stages 1 and 2 is on the children's need to select their own materials and to decide on the mathematics they need to solve the problems which arise. They are also expected to talk about their work, using the correct vocabulary, and to overcome difficulties by finding alternative ways of tackling their problems.

Case 1.2 This account illustrates the resourcefulness developed by 6-year-olds in one school I visited in which constructional work based on waste materials was considered to be of first importance.

Timothy had just finished painting a two-storey house he had made. It had curtains at every window and rugs on the floors. The symmetrical roof had four sloping faces and a square chimney was firmly fixed in place.

Timothy explained to me the difficulty he had experienced in fixing the chimney to the sloping roof. He had used sellotape first, but that had peeled off. At this stage, the teacher had asked all the children if they had watched their mother line a cake tin with greaseproof paper when making a fruit cake. Molly recalled seeing her mother snip round the edge of the bottom circle and fold the cut pieces up the sides. Molly's description made Timothy try this method on his chimney. He made short cuts all round the bottom of the chimney, then bent the flanges up and glued them to the sloping roof. Success!

Timothy also told me about the furniture he planned to make from stiff card. He had clearly thought about the project for a considerable time, and I was impressed with his work. On the spur of the moment, I asked him to draw what his house would look like from a helicopter just overhead. He immediately made a quick sketch (Fig. 1.1) which was full size and correct in every detail, even to the garden seat which had been attached under the eaves – perhaps because there was nowhere else to attach it?

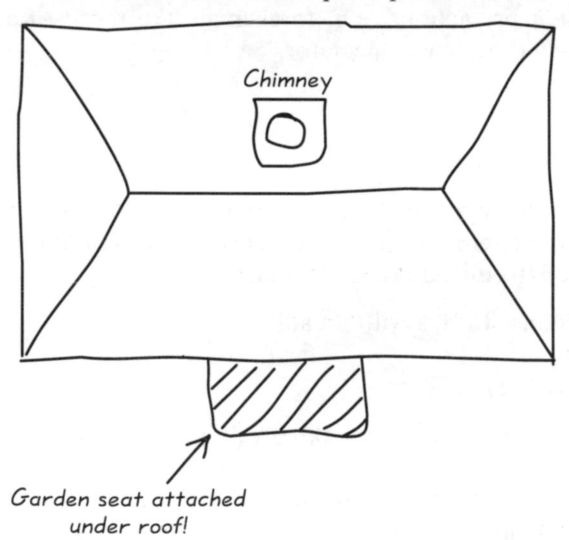

Figure 1.1 *House view from above*

I wondered if I had stumbled upon a mathematical genius. Remembering Piaget's experiments with young children, I then asked Timothy to draw a nearby table as though from a helicopter overhead. He drew a rectangle, this time in proportion, but then drew a short projection at each corner. My face must have shown disappointment, because Timothy said: 'You stand on the table and see what you see.' I did, and found that Timothy was right – the table was a stacking table and I could see clearly the short projections.

By now, other children were interested and wanted to join in. Timothy's table drawing had been turned over – out of their sight – so I asked them all to draw the table as though they were in a helicopter just overhead. They all drew a rectangle with a short projection at each corner!

When I showed the sketches to the teacher she confessed that she had no idea of what I was trying to do with the children. However, when I explained, she agreed that the constructional work had developed unusual powers of observation for that age.

The next time I visited this particular school I found that Timothy had made a book of plans, one for each room in the house, although this was not the only example I saw of systematic recording.

MODELS AND MAKING PREDICTIONS

For Key Stage 1, the Programmes of Study also state that children should be able to:

- talk about the models they make and make predictions based on their experience;
- make up and tell stories about the models they make;
- make predictions when using balance scales.

It is interesting that the ability to use alternative approaches to overcome difficulties is a Key Stage 2 requirement – as is the ability to record work systematically.

Pupils are also expected to test their predictions, e.g. to estimate the result of a written calculation or of one to be carried out on a calculator.

INSET

For teachers at Key Stages 1 and 2, it is important that they too should experience **model-making**, using the varied materials enjoyed by young children. Some of the problems children encounter could be introduced and discussed, e.g.:

- making and attaching a roof to a house (a box without a lid);
- attaching wheels to a vehicle; and
- using scraps of material to best advantage.

Teachers who have had practical experience of model-making will be able to suggest more problems themselves.

Prediction is another topic which may be unfamiliar to some teachers, e.g. predicting which of two objects is heavier, by first holding the objects one in each hand. They will also need to practise **estimating** answers to calculations, such as the multiplication of 37×42. The teacher should recognize this as greater than 30×40 (or 1200), and less than 40×45 or 1800.

Division (such as $158 \div 42$) is more difficult and estimations (such as $160 \div 40 \Rightarrow 4$) would not be made until Level 4. However, teachers should consider whether $160 \div 40$ is more or less than $160 \div 42$. (Since 42 is greater than 40, $160 \div 40$ is more than $160 \div 42$.)

Chapter 2

Number

> Pupils should understand and use number including estimation and approximation, interpreting results and checking for reasonableness.

This is the general statement for the Programme of Study.

INTRODUCTION

During the past ten years, powerful influences have affected the teaching of number throughout the primary school age range: 5 to 11 years. Today, teachers encourage their pupils to use electronic calculators for lengthy calculations, and to explore number patterns. Moreover, most children will have used computers for a variety of investigations.

There is no doubt about the attractiveness of calculators (and computers) for all children. They become very independent when using calculators, and are keen to explore patterns. They do not want to be shown different processes; they want to be left on their own to make their own discoveries. Recently, in the CAN project (directed by the late Hilary Shuard at Homerton College, Cambridge), some young children were introduced to calculators before they began a traditional programme on number – with interesting results. After some time, these children showed a preference for using their own mental or written calculations, saying that it was often as quick to think out their own methods as to put the information into the calculator!

On the other hand, a greater degree of mental competence in the four operations (addition, subtraction, multiplication and division) is now expected of children – who are given every opportunity to develop their own methods, and in general to be creative.

EARLY STAGES IN NUMBER

Making friends with numbers plays an important part in the education of young children (those at Level 1) if mathematics is to develop into a subject which is much enjoyed at a later stage. Unfortunately, for many parents, and even for some teachers, 'doing sums' is still the only indication that the child is engaged in mathematics. What is not so easily accepted is that when a child writes down an apparently simple sum, such as $8 + 7 = 15$, it involves several concepts and skills:

- matching two objects one-to-one,
- conservation of number,
- place value,
- the mathematical symbols of addition (+) and equality (=).

At this stage, **matching objects one-to-one** means pairing the members of two sets, e.g. mugs and spoons, to show whether or not there are the same number of objects in both sets. This concept of matching precedes the skill of being able to count meaningfully.

For Level 1, the list of activities suggested in the Programmes of Study includes:

- counting, reading, writing and ordering numbers to at least 10;
- learning that the size of a set is given by the last number in the count;
- understanding the language associated with number – more, fewer, the same;
- understanding conservation of number;
- making a sensible estimate of a number of objects up to 10.

For the second strand, pupils are expected to use addition and subtraction, with numbers no greater than 10, in the context of real objects.

So the children are learning to count the number of objects in a set soon after they first come to school, at about 5 years old. However, before they can understand what they are doing, they need to understand **conservation of number**:

> that the number of objects in a set remains the same however the objects in the set are arranged,

e.g. close together or far apart. Indeed, it is unprofitable to allow children to add two numbers before they are quite convinced, through many experiences, that these numbers do remain constant. In fact, it is the lingering doubt in a child's mind that often prompts him or her to continue 'counting in ones' long after they appear able to do written addition. Many children, when they have to add 8 and 7 together, are unable to count on from 8 but must go back to the beginning and start, 1, 2, 3, etc. Later on, they may say with confidence,

$$8 + 7 = 15 \quad \text{because it is} \quad 2 \times 8 - 1 \quad \text{or} \quad 2 \times 7 + 1$$

The concept of conservation of number is easily assessed, for example, as in the classroom activity given here.

Activity 2.1 Give the child a set of cups and a set of saucers (with different numbers, say 5 cups and 7 saucers). Ask the child to find out if there is a cup for every saucer. When he finds that there are fewer cups, let him see you remove the 2 extra saucers. Now arrange the cups in a cluster but leave the saucers spread out. Ask the child whether he thinks there is now a cup for every saucer or more cups or more saucers. (Sometimes use the word 'fewer'.) If he says, 'I know there are the same number of cups as saucers because I matched them', then the child has indeed acquired the concept of conservation of number. If, however, a child hesitates, then he needs further matching experiences.

Once children have acquired the concept of conservation of number, and can match the objects in one set with those in another set, and know the number names in order as far as 6, they can begin to count the number of objects in small collections by matching each object in turn to a number name said in order. Gradually children realize that **counting** is matching a number name (in order) to each object in the set. The following example shows how to organize progression in counting.

Case 2.1 The teacher in a rural school gave the eight children new to the school a daily activity in counting. To help them to learn the number names in order, she taught them number rhymes and games (*Number Rhymes and Finger Games* by Wrigley, published by Warne) which they enjoyed every day.

She provided ten small polystyrene trays, ten of each of ten different attractive plastic toys, and ten cards. Each card had a number symbol on one side and the matching number name on the other side. She began by introducing 'one', or '1', and put one toy car in the tray marked '1'. Later that day she found that all the children in the group knew that they had two hands, two feet, two eyes and two ears, so she showed them the symbol '2', and Ann chose two plastic cats to go in the tray with the symbol '2'. During the day, the children put two hands together and said '2'. Each day, a different child filled the trays with first one object and then two objects, and labelled the trays with the correct number symbol or number name. At the end of the day, the same child counted the objects as he returned them to their set.

In the following week, the teacher introduced '3', put three donkeys in a tray labelled '3' and asked the children to arrange the three trays in order, after making sure that they remembered the addition and subtraction facts about '1' and '2' and the new number '3'.

Every day the teacher gave the children new sets to count. For example, '4' is the number of children working in the book corner and in the water corner, and '5' is the number of windows in the classroom. Soon they found their own collections to count. As the teacher introduced each new counting number, she always made sure that they remembered the number facts they had learned previously.

One day, when they could count to 10, recognize the symbols for the numbers 1 to 10, and fill the trays correctly, she added an empty tray at the beginning. She asked them if there was a number which would fit in the empty tray. They could not think of one, so the teacher hid zero '0' among the cards 1 to 10, where Pat found it next day and put it, triumphantly, in the empty tray at the beginning.

Sometimes, the teacher jumbled the cards and asked the children to arrange them in order before they filled the trays. At other times, she jumbled the cards and asked the children to fill the trays before they arranged them in number order. Every day, the different groups of children talked about the work they were doing. Sometimes, they made number patterns with small objects and described the patterns in as many different ways as possible.

Discussing number patterns is an important activity in memorizing number facts.

LEVEL 2: ADDITION AND SUBTRACTION

Addition is the inverse operation of subtraction: subtraction is the inverse operation of addition, or we can say: subtraction undoes addition and addition undoes subtraction, e.g.

$5 + 3 = 8$
$3 + 5 = 8$
$8 - 5 = 3$
$8 - 3 = 5$

It is important that children should have many experiences of problem situations which require addition and of those which require subtraction for their solution.

Addition

Suggest that the children make collections of one kind of object, e.g. pebbles, shells, postcards, flowers, etc. Ask questions such as:

- Did you add any . . . to your collection today?
- How many did you add?
- Can you put the collection in a pattern to show how many there are?
- Are there more than five?
- Are there more than you can count?

Examples of addition are easily found:

- A trolley at a supermarket is filled as the family moves round the aisles, adding more things.
- A child sees the amount of milk in his glass increase as his mother pours it out for him.
- As water is run for his bath, the amount increases as more water is added.
- As the child puts his bricks in one pan of balance scales, the pan goes down (because it is heavy).
- As more air is blown (added to) a balloon, the balloon gets bigger.

There are many other examples of addition in the classroom and about the home and school.

For each situation, it is important to introduce the correct vocabulary and to make sure that the children use this themselves:

add *increase* *more* *total* *sum* *altogether* *addition*

Subtraction

There are two types of problems which are solved by subtraction, and these can be expressed by using three different language patterns. The first type of problem involves taking away some objects from a set, or a smaller amount from a quantity, and is the reverse of addition. Indeed, all the situations which serve as starting points for addition, when reversed, become subtraction situations – of this first type, termed *take away*.

- The children's collections become smaller as they lose items, or give them away.
- They notice that the number of objects in their collections becomes fewer.
- The child sees the level of milk in his glass falling and gradually disappearing as he drinks it.
- He sees the bath water draining away when he removes the plug.
- He watches the scale pan rise and sees balance restored as he removes bricks from one pan.

Teachers need to find many opportunities for children to experience problems which require this first type of subtraction for their solution. As the children finish each problem, they should be asked to show how much/many is/are left.

Activity 2.2 Ask the children to empty the water from the water bath, then ask them how they did this.

> I got a jug. I filled it with water from the bath. I poured the water down the plug hole. Everyone helped. We took water out until we could carry the bath and pour the rest down the sink. There was none left.

Activity 2.3 Another recurring subtraction activity involves the class shop, and real money. There are many activities in which the children buy one or more items, and receive change, at first from 5p, and later on, from 10p, then from 20p and from 50p. The change is counted out by the shopkeeper:

> You bought a tomato for 3p, and 2p makes 5p.

The children record:

$3 + 2 = 5$ and also $5p - 3p = 2p$

The second type of subtraction problem to be solved is the comparison of the numbers in two sets, or of two quantities. So the starting point for this type of subtraction is two sets: two heights, capacities, areas, or 'weights'. The questions to be asked are of the type:

- How much/many more?
- How many fewer?
- How much less?
- How much longer/shorter is the doll's bed than the teddy bear?
- Is the bed long enough?
- Are there more cups or more saucers in the cupboard?
- How many more cups are there?

A third language pattern can be applied to the two-set problem. It is allocated to Level 2 but some children will be ready to use this alternative language pattern before then, although others find this pattern more difficult. It is:

- Find the *difference between* your height and your brother's.
- Find the *difference between* the cost of a packet of nuts and a cornet.
- Throw two dice 20 times. Record the *difference between* the scores each time. Which difference occurred most often? Which difference occurred least often?

The vocabulary needed for subtraction is more complex than that for addition, and involves many more terms:

subtraction *take away* *spend* *spent* *change*
how much is *left*? how many are *left*?
minus *comparison* *more* *fewer*
what is the *difference* in number? in length? etc.

Children need to encounter many real-life problems in addition and subtraction at regular intervals if they are to become familiar with these operations, recognize them at sight, solve the problems and express themselves using the appropriate vocabulary. A class shop provides many opportunities for carrying out problems involving addition and subtraction (and multiplication and division) using real money, but also gives an opportunity to introduce the 'shopkeeper's addition' method of subtraction. (This is a very useful method later on, when the children are presented with more difficult subtraction, e.g. $521 - 378$, which is more easily solved by 'adding on'!) For example, when giving change from 50p after 28p has been spent, the shopkeeper says: '28p and 2p makes 30p, and 20p makes 50p.' (Later on, the children can choose to use a number line for these harder examples.)

DIVISION AND MULTIPLICATION

Addition and subtraction do lead easily into multiplication and division, but multiplication and division problems do not appear in the National Curriculum until Level 3. However, it is important with most children to make a start at Level 2 on the problems which require multiplication and division for their solution.

Division is the inverse operation of multiplication; multiplication is the inverse operation of division, or we can say, division undoes multiplication, and multiplication undoes division, e.g.

$7 \times 4 = 28$
$4 \times 7 = 28$
$28 \div 4 = 7$
$28 \div 7 = 4$

In daily life, division situations occur far more often than multiplication situations; we therefore begin with division.

Division

There are two types of problem which are solved by division (sharing, and subtraction of equal sets), and these give rise to three different language patterns.

Sharing

Children are usually involved in sharing experiences at home, with their siblings, before they come to school. These experiences include sharing fruit, a bottle of pop, a length of ribbon, a lump of clay or a bag of raisins with one or more siblings. It is important to notice how children solve these problems. The teacher should try to avoid showing them what to do – they may not be ready for these activities yet. At school, there are certain to be one or two children who can share successfully. The materials needed for this activity are some identical, transparent containers, and some top-pan balance scales (for halving clay, and for comparing guessed 'halves'). It is in sharing activities of all types that young children first experience fractions and learn the language of one-half, a quarter, three-quarters, etc.

Activity 2.4 Ask the children to share between themselves and a friend: a length of ribbon, a glass of water, a sheet of paper. Watch carefully to see how they do this. Ask them to describe their method.

Halving and quartering are some of the different forms of sharing, and similar language patterns are used.

Subtraction of equal sets

In the National Curriculum the subtraction of equal sets is referred to as 'equal partitions in'.

The following activity can be tried with young children who are still at the nursery stage, provided that they know how many 'two' are.

Activity 2.5 Show a group of young children a bag (or plate) of shells, and ask: 'Do you think there are enough shells in this bag (or on this plate) for everyone in this group to have two each?' The usual response is: 'Let's try and see.' (Always make sure that there are indeed enough shells in the bag for everyone in the group to have their 2!) So the bag of shells is passed round to each child in turn. As each takes his/her shells from the bag he/she is asked: 'What have you done?' They usually say: 'I've taken my 2.' (Notice that they are using the language pattern of subtraction.)

When the last child has taken two shells, the teacher asks: 'Are there any shells left?' The child looks, counts and tells us.

When the children have completed similar activities several times, they should be asked if anyone can tell how many shells there were in the bag to start with. Sometimes they solve this by counting round the group in twos and adding the remainder. (This is, of course, a multiplication problem: the addition of equal sets.) Others solve the problem by multiplying the number in the group by 2 and adding the remainder.

Children need a variety of activities of this kind at frequent intervals, and should be questioned to find out whether they recognize each activity as the subtraction of equal sets.

Activity 2.6 Often such tasks involve an element of **estimation**. For example, ask: 'How many glasses can you fill from this jug?' Ask them to estimate first, and then to check. It helps if there is a line near the top of the glass so that all glasses can be filled to the same level.

Then give each group a length of ribbon and ask them how many (equal) hair-ribbons they can cut from the long piece. First, they will need to decide on a suitable length for a hair-ribbon. Then ask them to estimate how many ribbons of this length they think they could cut from the long piece. Then, within their groups, they experiment to check their estimate. To follow up, the teacher asks:

- Which group's estimate was nearest?
- Did you have any ribbon left?
- Is there enough left over for another ribbon?

Then, finally, they are asked: 'Are all your hair-ribbons of the same length?'

The vocabulary of division includes the following:

equal shares *sharing*
Is there any *left over*?
How many *equal shares* are there?
Is there a *remainder*?
divide by *division*

Multiplication

As in division problems, there are two different types of multiplication problems which correspond to the two types of division. (As an exercise, try to identify these before reading on.)

Multiplication is the inverse operation of division. The first type of multiplication problem which children meet is the **addition of equal sets**, the reverse of the subtraction of equal sets. For example, if we are buying one kiwi fruit costing 12p for each member of a family of 6, it is far quicker to find the cost by multiplying 12p by 6 than it is by adding 12 six times. We learn our multiplication tables because multiplication is a quicker method than addition.

Activity 2.7
1) Present the children with the task of working out this greengrocery bill for the Smith family:

5 kiwi fruits at 16p each
4 pounds of potatoes at 18p a pound
3 pounds of cooking apples at 38p a pound
6 pounds of Cox's apples at 49p a pound

Ask them to calculate how much change there would be from £10.
Then ask the children to make out their own bills from the prices their mother pays – or let them make a class greengrocery shop.

2) Other problems in multiplication which arise naturally are on the sports field. Suppose that 6 children run 4 laps of 60 metres each.

- How far does each run?
- How far do they run altogether?
- How far short of 2 kilometres is this?

It is important to emphasize that multiplication is a quick way of adding equal sets. At the same time, it is important to encourage the children to begin to memorize the multiplication tables – first, the doubling table (multiplication by 2); then multiplication by 10; then halving for multiplication by 5.

The second type of problem solved by multiplication is **magnification**, the reverse of the sharing aspect of division.

- Find a length of string 4 times as long as this.
- Find a jug which holds six glassfuls.
- Find a lump of clay twice as heavy as this piece (provide top-pan balance scales).
- Find a number ten times as many as 13.

For this, the vocabulary includes:

multiply by *add equal sets* *magnification*
– times as many as –

INSET

Although all teachers can use the four operations, it is still important to include them in the early stages of the INSET progamme. How they approach the subject at this lowest level can have great (good or bad) effects on the later development of their pupils.

- Discuss the problems and activities included in the previous sections on addition, subtraction, division and multiplication.
- Add to the list of activities.
- Make sure that everyone understands the mathematics involved within these operations, and their interconnections.

RECORDING THE FOUR OPERATIONS

The children's first descriptions of their activities in the four operations will be oral. It is important that they can use the correct vocabulary themselves. As new words are introduced, the teacher should provide pictures and a written version, so that the children gradually begin to make their own pictures and then a written record. It is particularly important not to introduce the **mathematical symbols** until the children have had many experiences of real situations, in number and in measuring, within the classroom and in the rest of the school. Then, when they are introduced, care should be taken with the wording used:

- the *addition* symbol '+' meaning 'put together and say how many';
- the *subtraction* symbol '–' meaning 'take away' or 'find the difference between';
- the *multiplication* symbol '×' meaning 'multiply by' (representing the total of a number of equal sets or magnification) or 'times as many/much';
- the *division* symbol '÷' meaning 'divide by' (representing sharing among, or finding how many equal sets in).

Division and 'remainders'

There is a distinction to be made between the equal sharing of quantities and sharing a number of objects.

When children are asked to share a small jug of lemonade among three children (providing three identical glasses), they will finish the jug, matching liquid levels and leaving nothing in the jug. Sharing a length of ribbon is often done by folding (this is a little more difficult if the ribbon is to be shared among three) and, once again, no ribbon is left.

When asked to find half or a quarter of a sheet of paper, this they do by folding, and (again) leaving no remainder. Similarly, if asked to halve a lump of clay, the children may well use balance scales (if these are readily available) and divide up all the clay.

However, when children are asked to share, say, a collection of shells or pebbles, it is likely that there will be a number of objects left over. In this event, it is better for them to give the remainder as a number.

By Level 3, most children should be clear about the two division problems:

- sharing the number or quantity among a number of people (finding how many/much there are/is in each share);
- finding how many equal shares can be obtained from the given number or quantity.

Thus $12 \div 4 = 3$ can represent:

12 divided into 4 equal sets (with 3 in each)
or
12 divided into sets of 4 (3 equal sets)

Numerical methods

We cannot really discuss these until we have considered place value; this is begun in Level 2 and continued through Level 3. At Level 2 the Programmes of Study other than those associated with the operations of addition and subtraction state:

- reading, writing and ordering numbers, first to 10 and eventually to 1000, using the knowledge that the tens digit indicates the number of tens;
- solving whole-number problems involving addition and subtraction, including money. Using coins in simple contexts.

Three other activities involve measurement:

- using non-standard measures for comparing objects and events;
- recognizing the need for standard units;
- learning and using the common standard units of length, capacity, 'weight' and time.

In Chapter 3 we discuss progression in measurement, Levels 1 to 6, but here we begin with place value.

PLACE VALUE

Our counting system and the way we write our numbers is well-structured and economical. We need only ten symbols (or digits): 0, 1, 2, 3, . . . 9, to write a number, since the magnitude of each digit depends on its position. For example, in forty-seven, written '47', the '4' is worth 40 (4 tens) and the '7' is worth 7 ones or units. In '138', the '1' is worth 1 hundred, the '3' is worth 30, and the '8' is 8 ones or units. Teachers are all familiar with the sequence:

1000 100 10 1

in which each subsequent number to the left is ten times the previous number. So children have to learn to group in tens when they are counting, to change each set of ten ones for 1 ten-stick, and to move this ten-stick one place to the left.

It is thought that a long, long time ago, men began to count in tens because we have ten fingers. People who lived near the equator and were barefoot counted in twenties! (Twenty was called 'Man finished'.)

A number system based on place value involves counting in sets of a given number, usually ten, then moving each completed set one place to the left. Of course, it is not essential to count in tens. Indeed, it is valuable to introduce young children to smaller counting numbers – to give them adequate practice in the mechanics of place value – organizing in sets (e.g. of four) and moving each completed set one place to the left. Counting in 3s and 4s takes far less time than counting in 10s, and children can have more practice in the same time. Later on, their teacher can help them to generalize and use 10 as *the* counting number.

Teachers will need to decide for themselves when individual children are ready for the place-value activities which follow. The progressive set of activities provided is condensed, but includes one example from each level of development. However, the children will need a number of comparable activities at every stage – sometimes for a short period daily over several weeks – until they are successful, use the vocabulary correctly and have gained confidence.

All teachers will have their own carefully structured place-value activities – structured because our counting system itself is highly structured! Some will use Dienes' Multibase Arithmetic Blocks consisting of units, bars called 'longs', square cuboids called 'flats', and cubes called 'blocks'. These are in bases 3, 4, 5, 6 and finally 10. The material can be used informally for the four operations, using base 4 first. Later on, the material will probably be used in a formal way with each operation recorded carefully.

Dice games such as 'build a cube' and 'break a cube' provide valuable experience of changing smaller units for the next higher unit in an addition problem, and changing larger units for the next smaller unit in subtraction problems. This material can also be used to illustrate the different aspects of subtraction, multiplication and division. (Note to Mathematics Co-ordinator: if your school is not already familiar with this material, aim to use it in your INSET group and discuss the outcomes.) Obviously, it is important to vary the equipment used.

INSET

For each activity suggested, try them in your INSET group and discuss their merits. Those considered to be of value for the children to be taught, can then be incorporated in the teaching programme.

Activity 2.8 Start by taking a set of counting material, numbering fewer than 10. Group in 2s and state your score (e.g. for 8 it is 4 sets of 2 and zero). Then group in sets of 3 and again state your score. Continue with sets of 5. Repeat this activity using other material, e.g. identical sticks (3 make a triangle, 4 a square, and so on).

Activity 2.9 You each use an adapted egg box (Fig. 2.1) and small counting material: pebbles, shells or buttons. Start with a set of 7 pebbles in the 'ones' compartment of the egg box (the lid, on the right).

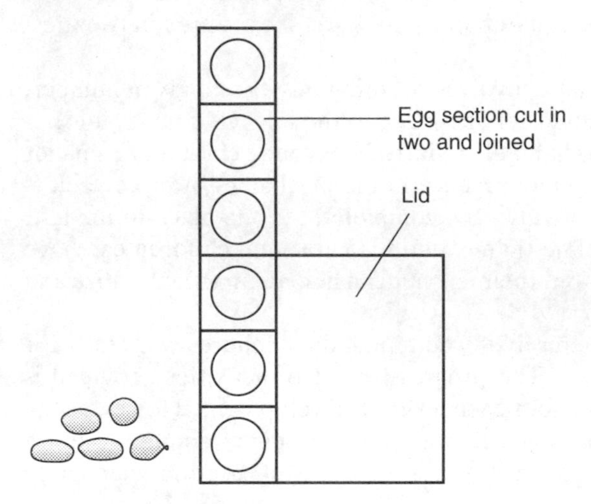

Figure 2.1 *Converted egg box*

- How many sets of 3 do you have? (Did you remember to move each set, as you completed it, to an egg space on the left?)
- What is your score? (2 sets of 3, and 1)
- Begin again.
- How many sets of 4 do you have? (1 set of 4, and 3)
- How many sets of 5?
- Use different starting numbers.

Activity 2.10 This activity depends on the limitation of the number of sets according to the counting number. The leader of the group chooses a number, which must never be mentioned during the game. Suppose this number is four, and the group leader decides to call it 'fish' (because fish has the same number of letters?). The game starts with all pebbles outside the egg box. Every time the leader rings a bell, the players in the rest of the group each put one pebble into the ones compartment, and the leader chooses one player and asks: 'What is your score?' The correct score is 'zero fish one'. The leader rings the bell a second time, and asks a different player the score when the players have added another pebble. After ringing the bell a third time the leader asks yet another player for the score. This time the answer should be 'zero fish 3 ones'. Then the leader asks: 'What will the score be when I ring the bell once more?' If the answer is 'fish ones', the leader asks the players what they should do next. They should suggest moving the fish ones to an egg space on the left. The correct description is now '1 fish 0 ones'. The leader rings the bell and checks that all the players have the correct score.

The leader continues to ring the bell, gives the players time to move, then asks another player for the score. The leader should pause before each new set of ones is complete to make sure that all the players remember what to do at the next bell. When the score is '3 fish 3 ones' the leader checks with the players what they will do if the bell rings again. The players are given time to say that they would move another fish ones to the left. That would make 'fish fish', which would have to be moved yet another place to the left. Unfortunately, by now there is no more room in the egg box!

This activity can then be tried using the number 5 – perhaps called pent.

- Is the activity easier this time?
- When did you have to stop because there was no more space?
- Was this when you reached '4 pents 4 ones'?
- What would the next unit have been? (1 pent pent)

It is important for teachers to appreciate the value of these addition activities, before trying them with the children they teach. Also, some children will need regular, short practice sessions over a long period. For variety, choose a different material (interlocking cubes are useful) and change the counting number. Sometimes a child could be given the role of the bell-ringer.

Next we consider some subtraction activities.

Activity 2.11 You need adapted egg boxes and counting material, as before. To signal subtraction, the leader also needs a triangle or shaker. Start with the material outside the egg box, and play the addition game for counting number 5 until the score is '4 pents 4 ones'. The leader then explains the new game: every time the triangle is struck the players remove one pebble from the ones compartment of their egg boxes.

The leader asks a different player each time to announce the score ('4 pent 3 ones' the first time). When the score is '4 pents 0 ones', the leader asks: 'What should you do at the next strike?' (Answer: change a pent for 5 ones and put them in the ones compartment on the right.) Remove 1 pebble. The score is now '3 pents 4 ones'. The leader continues to strike the triangle, asking the score and pausing to ask what action should be taken at each strike when there are zero ones.

Now, this activity can be repeated using the counting numbers 4, 6 and 3. As an alternative to the egg boxes and pebbles, use interlocking cubes and start with a score of '3 quad 3 ones'. Other starting numbers can be used. Sometimes ask a player to act as striker. It is also possible to mix addition with subtraction.

Again, it is important for teachers to appreciate the value of the activity before using it with the children. It is essential to vary the materials used, and to give the children time to become familiar with the new pattern of working. Some teachers prefer to omit the previous exercises, and go straight to 10 as their base number. We use ten as the counting number for the remainder of the activities in this section.

Activity 2.12 The children work in pairs and need interlocking cubes in one colour, a die for each pair of children and a place value sheet for each child. They throw the die alternately and take the score in cubes. The thrower must then wait for the opponent to say what has to be done. For example, player A with score 7 throws 4. Player B says:

- Take 4 cubes.
- Add 3 to your 7 to make 1 ten-stick.
- Move it to the left (tens column).
- What is your score? (Answer: 1 ten 1).

One advantage of each player telling the opponent what to do is that the teacher can check that each child uses the correct language. The winner is the first player to reach 3 tens or more. Next time the goal should be exactly 3 tens. Variations on this activity extend the number range used.

- Throw two dice and add the scores, recording with interlocking cubes. The winner is the first to reach 50.
- Throw two dice. Each player can choose any operation at each throw. He tells his choice to his opponent, who then gives him instructions. The winner is the first player to reach exactly 100.
- First to zero. Two players each start with three ten-sticks. They throw the die in turn. At each throw the player subtracts his score from his total. As before, he must wait for his opponent to tell him what to do. If player A's first score is 4, his opponent says:
  - Change a ten-stick for 10 ones and move these to the ones column.
  - Remove 4 ones.
  - What is your score? (Answer: 2 tens 6).

The first player to reach 0 is the winner.

Abacus games (Level 3)

For these activities, which involve changing 10 unit beads for 1 ten-bead, each child in a pair needs a 3-spike abacus, 10 red (1), 10 blue (10) and 1 green (100) beads to fit the

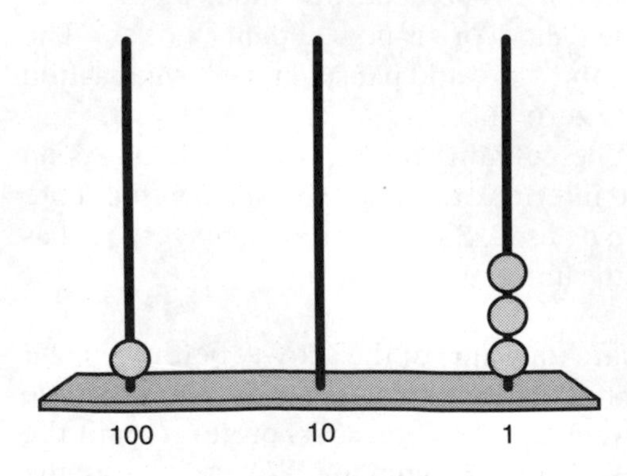

Figure 2.2 *Abacus*

abacus, two dice and a shaker (Fig. 2.2). The abacus can be made from an unused strip of Plasticine and 3 knitting needles.

Activity 2.13 The players throw two dice in turn and add the scores. The opponent has to tell the thrower what to do. For example: Player A, score 8, throws 5 + 4. Player B says:

- Take 9 red beads. Your score was 8 units.
- Put 2 with the 8 red beads, making 10 red beads.
- Change 10 red for 1 blue bead and put it on the ten spike.
- Put the 7 remaining red beads on the ones spike.
- What is your score? (Answer: 10 + 7 = 17).

Continue the game. The winner is the first to place a bead on the 100 spike. As a variation the players may opt to use any operation at each throw.

Activity 2.14 Two players each start with a score of 100 (1 green bead on the left-hand 100 spike). At each throw the player subtracts the score from the score on the abacus. To begin the players must change the 100 bead for 10 blue beads, then change 1 blue bead for 10 red beads, and put the beads on the correct spikes: 0 green, 9 blue, 10 red. Player A then throws 2 dice and adds the scores, e.g. 5 + 3 = 8. He removes 8 red beads from the 'red' right-hand spike. Player B asks him for his score (9 tens and 2). Player B then has his turn and the game continues. The winner is the first player to reach zero.

To check their understanding of the processes involved, players should be asked under what circumstances would they need to change 2 blue beads to begin the game? Teachers can adapt this idea to make up their own versions of the game. It would be possible to repeat Activities 2.13 and 2.14 using beads of one colour only. When teachers are quite familiar with these games they can try them with the children they teach.

Also the abacus with beads of one colour can be used to represent a variety of numbers for the children to identify: for example, 101, 503, 495, etc. Or the children could be asked to represent various numbers, using beads of one colour. It is also useful to give them simple addition and subtraction on an abacus. The following activity incorporates an element of **estimation** and should be a daily activity for a period of time.

Activity 2.15 Prepare collections of small objects (pebbles, buttons, seeds) for groups of children to estimate the number. Each records his estimate (which can be based on a set of 10). The group then counts the collection, patterning in sets of 10 and the remainder, for the teacher to check the number at a glance.

Place-value games at Levels 1, 2 and 3 can provide an enjoyable learning experience. The children play in pairs. Each child needs a place-value sheet, folded in half along the long edge and labelled 10s 1s, and one set of cards, each showing one of the digits 1, 2, . . . , 9. Each card set should be shuffled and placed face down in front of each of the two players.

Activity 2.16 Each child takes two cards, one at a time and in turns, to make a 2-digit number; the aim is to form a higher number than their opponent. Player A takes his top card, shows it and decides whether to put it in the 10s or in the 1s column. Player B then takes the next top card and decides where to put it on his sheet. (Once placed, cards must not be moved.) Player A then takes his second card which has to be placed in the empty column. Player B does likewise.

Each player says his number aloud. The winner (with the higher number) scores 1 point (Fig. 2.3). The children can play five separate games. This game offers many extensions/variations, e.g. letting them play five games in which the player with the lower number wins. Equally, bonus points may be offered to the loser if he can say by how many he has lost. (Number lines and ten-sticks to fit should be available.) This place-value game can also be adapted for children of different ages. For example, three or more cards can be used, the cards being placed in either column and the numbers added as necessary. This time the player whose score is nearest to 100 is the winner. An extra point is scored if at any stage a player can say how many more points he still needs to score 100.

At a later stage, include zero in their packs. Finally, each player takes six cards in all, one by one, and places each in turn in either column as before. Again, the winner is the player with the score nearest to 100. When a player scores over 100, he loses the game immediately. He scores an extra point if his score is exactly 100. A specimen game is given in Figure 2.4.

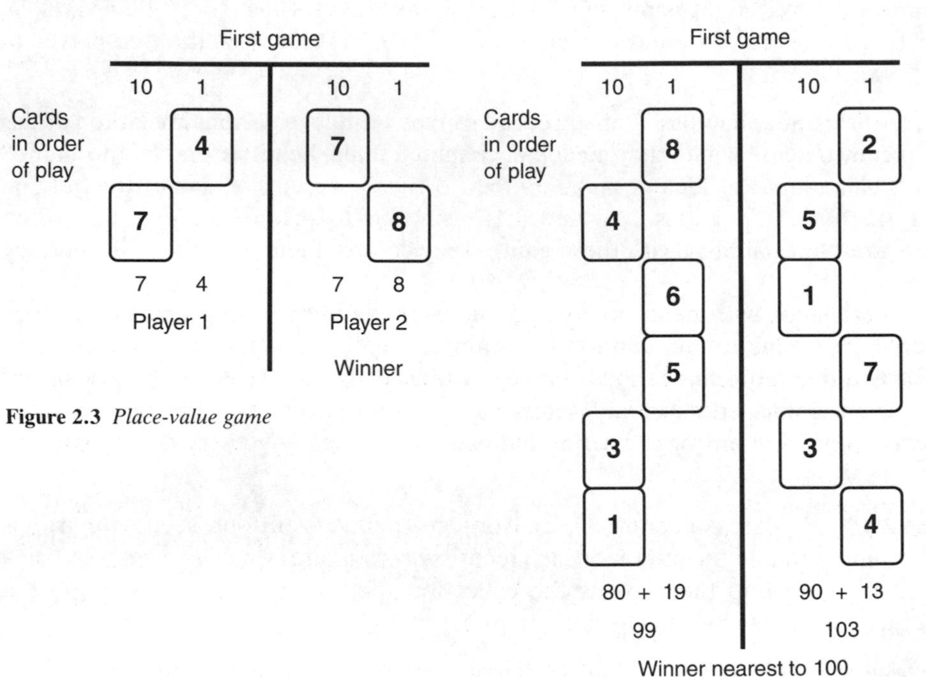

Figure 2.3 *Place-value game*

Figure 2.4 *Another place-value game*

Assessment

The work at Levels 1 and 2 has comprised mainly practical activities. As the children carry out these activities, talk about them and record their findings, the teacher will be able to assess their understanding. However, it is also important to check whether individual children understand their own written records, e.g. 57 represents 5 tens and 7 ones, etc.

Written calculations will provide teachers with an important means of assessment. Teachers first need to make sure that children are able to work through the following examples and to make a written record of each.

- Identify and record 2- and 3-digit numbers set out on a 3-spike abacus, e.g. 358, 709, 600, 014.
- Use a 3-spike abacus to set out numbers dictated orally.
- Arrange three or four different numbers in size/order (smallest to largest).
- Write six different numbers using the same three digits, arrange these numbers in size order and find the difference between the largest and the smallest. For example, 2, 7, 5 produces 257, 275, 527, 572, 725, 752, with a difference of 495.

1. Identify, in a display of numbers, the values of a single digit, e.g. 8 in 982, 8045, 1807.
2. Tell how many different numbers they can make with the digits used in 902.

Children working in pairs can be assessed while they work on the abacus. They talk as they work, describing what they are doing, and then make a written record.

ESSENTIAL NUMBER KNOWLEDGE

From the earliest stages, young children are gradually building up a systematic number knowledge, first of addition and subtraction facts, then of multiplication and division facts. This knowledge is vital if children are eventually to be able to solve the problems they come across – in their heads, in writing or by using calculators or computers (see Chapters 8 and 9). Moreover, teachers and pupils need to keep up-to-date individual assessments of the extent of their knowledge.

Addition squares serve many purposes at this level, not least to record the addition facts for pairs of numbers from the set 0 to 10 (see Fig. 2.5). Children can be presented with a complete table and asked to look for the number patterns and describe these, including the patterns on the two diagonals. Can they explain why the doubles of numbers 0 to 10 occur on the major diagonal? How else can they describe the numbers on this diagonal? Sometimes they can be asked to analyse the table – listing the number of times each total occurs in the table for pairs of numbers (a) 1 to 6, (b) 0 to 10.

It is important not to make an issue of learning these facts in case any child becomes over-anxious, but they should be given regular practice until they respond promptly and accurately. They should know the addition (and subtraction) facts of all the pairs of numbers from 1 to 6 (36 facts in all) before extending the set to 0 to 10 (121 facts).

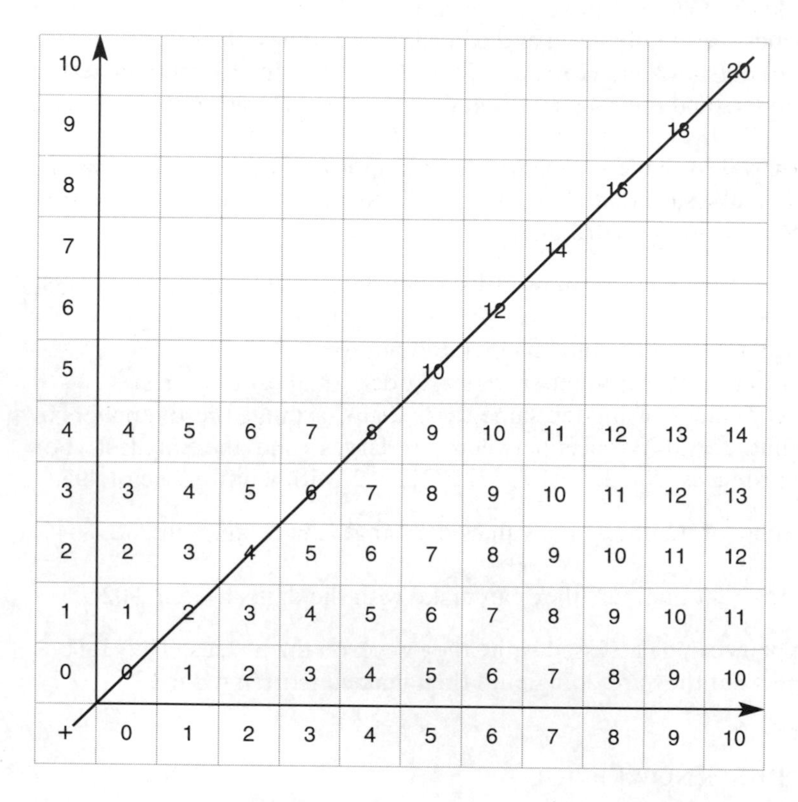

Figure 2.5 *Addition table*

Activities which involve addition (and later, subtraction) of the scores on two dice can provide purposeful practice. Ask the pupils to collect the aggregate results of the class for each of the totals 2, 3, to 12 and to display them in the classroom. Later on, these experimental results can be compared with the occurrence of the totals in the addition table for numbers 1 to 6.

Essential addition facts

The following list itemizes the essential addition skills which every child needs.

- Adding 1 to numbers 0 to 10: $0 + 1 = 1$; $1 + 1 = 2$; $2 + 1 = 3$; and so on.
- Adding 2 to numbers 0 to 10 – making sure also that the pupils have a practical method of distinguishing between odd and even numbers: $0 + 2 = 2$; $1 + 2 = 3$; and so on.
- Knowing pairs of numbers whose sum is 10: $0 + 10 = 1 + 9 = 2 + 8 = \dots = 9 + 1 = 10 + 0$.

- Immediate response to addition of 10 to numbers 0 to 10 – having tens and units equipment available provides a short cut in helping children to discover these facts.
- Knowing the quick addition (and subtraction) of 9: $4 + 9 = 4 + 10 - 1$.
- Doubling numbers (also halving) up to 10 (and eventually to 50).
- Spotting near doubles, e.g. $8 + 9$, $7 + 6$, $6 + 5$, and so on. Ask pupils to tell you how they found out that $7 + 6 = 13$ $(2 \times 7 - 1)$ or $(2 \times 6 + 1)$?
- Adding number pairs differing by 2 e.g. $9 + 7 = 2 \times 8$. Try to obtain all these responses from the children by questioning.
- Addition (and subtraction) of zero.

The addition square can also be used for assessment purposes. Provide square centimetre paper and ask the pupils to prepare a blank addition chart for pairs of numbers from 0 to 10. (The pupils could prepare a duplicate for their own records.) Figure 2.6 shows one pupil's progress in the memorization of pairs of these addition facts. The square can be annotated to show progress.

- A tick at the end of row 1 shows that he can add 1 to numbers up to 10.
- The ticks in symmetrical positions show that he knows both $5 + 4$ and $4 + 5$.
- The ticks on the diagonal show that he knows the doubles of numbers 2 to 8 (and also 8×2).

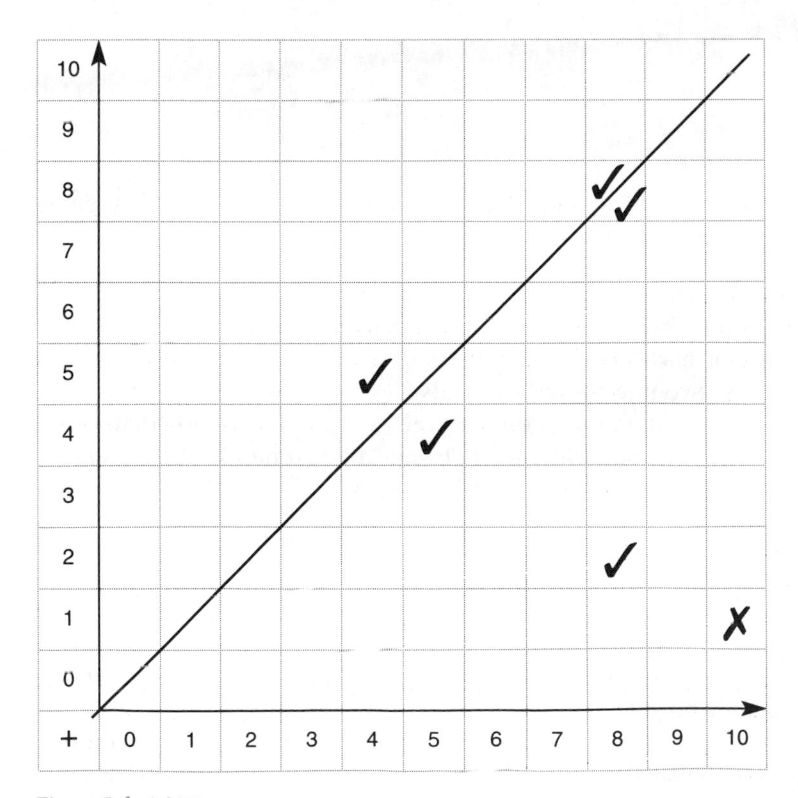

Figure 2.6 *Addition assessment*

In the early stages teachers should check progress of individuals at fortnightly intervals. A crossed tick in a square can indicate that a pupil has retained this knowledge for a term.

A multiplication square for numbers 0 to 10 can be used in a similar way to record knowledge of multiplication facts.

Subtraction

There are two forms of the subtraction table to consider for the numbers 0 to 10: teachers of younger children should make a **difference table**, always subtracting the smaller number from the larger, while teachers of Levels 3 and 4 should make a **subtraction table** to include negative numbers.

Activity 2.17 Ask the children to analyse the table (difference or subtraction according to their level) according to the number of times each difference occurs in the table. Then ask them to throw 2 dice and to record the difference of the scores at each throw. They can use calculators to compare the frequency of occurrence of the differences in the table (out of 36) with those from your experiment (out of the total number of throws).

The diffy game

Adults as well as pupils find subtraction harder than addition, and the diffy game provides interesting practice in subtraction.

- Draw a large diamond shape on a clean sheet of paper.
- Write a different number fewer than 10 at each vertex.
- Mark the midpoints of each edge approximately and at these points write the difference between the two numbers at the ends.
- Join the midpoints and put in the midpoints of the edges of the second diamond.
- At each new midpoint write the difference between the two numbers at its ends.
- Continue this process until you have to stop (Fig. 2.7).

The first question is: why do you have to stop? Try again, using two-digit numbers this time. What is the greatest number of sets of differences you took? Can this be related to the number of diamonds you drew? If you are interested try finding differences of numbers at the vertices of different shapes (a triangle?).

Having experimented with the diffy game with colleagues, try this activity with the pupils. Some children will be ready for this game at Level 2, and they can make attractive displays if they draw successive diamonds with crayons of different colours. Initially, keep to numbers fewer than 10, until the pupils have a confident knowledge of subtraction. When they can subtract tens and other 2-digit numbers, challenge them to choose starting numbers which might allow them to draw more than seven diamonds.

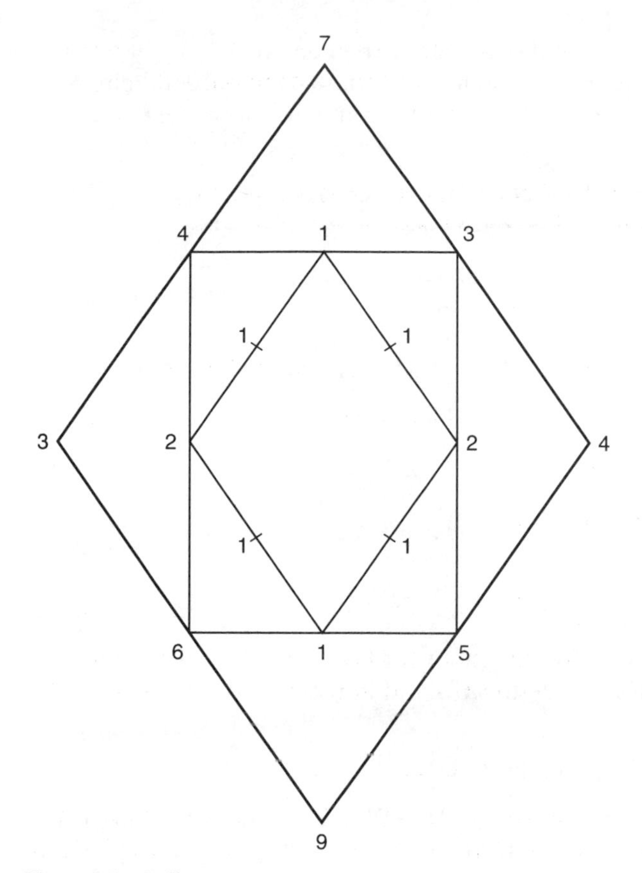

Figure 2.7 *Diffy game*

Using a number line

When extending the knowledge of addition and subtraction facts to 100 (Level 2), a number line can prove invaluable. Although a floor number line may have been used by your pupils from time to time, each pupil should now be given a metre strip of centimetre-squared paper, 2 cm wide, to prepare his or her own 100-unit number line.

It is best if they use pencil for the markings at first – there are sure to be some mistakes! The markings should appear on the lines (and not between them), with the extremities (0 and 100) written sideways to attach them to the correct lines. Then ask the pupils to label the halfway mark. (Do they fold the number line to find this point?) After this they label the tens (in black) and the fives (in red). They can cut an extra 10-strip if they need this.

Case 2.2 With number agility in mind (pupils should be able to add and subtract mentally two 2-digit numbers), I worked with a small group of 9-year-olds. I first tested their knowledge of simple addition (2-digit numbers without carrying, e.g. 37 + 22). Then I introduced pairs of 2-digit numbers with carrying, e.g. 48 + 24. Not all of them were successful but they had more fluency than I expected for a mixed group. Next, I tried some mental subtraction, e.g. 48 – 23. I had to repeat the numbers more than once before the pupils gave the correct answer.

When I switched to '75 − 57' there was panic and I knew they needed help. So, I gave out their number lines and they all found the two numbers on them (Fig. 2.8).

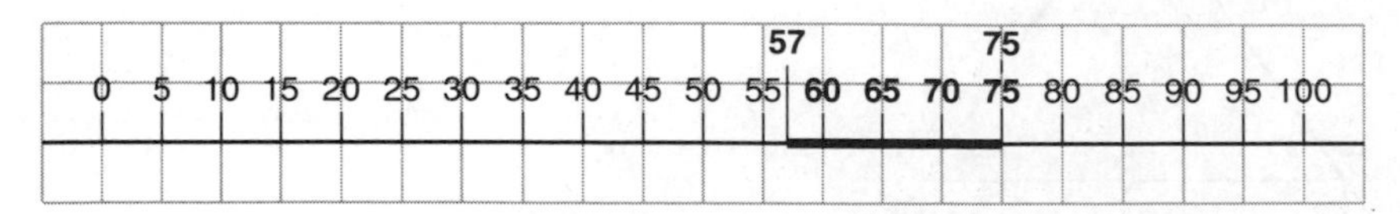

Figure 2.8 *Using a section of the number line to calculate 75 − 57*

Bianca suggested they find the distance between the two numbers. She counted:

> 57, and 3 to 60, 60 to 70 is 10, and 5 to 75, that's 18 total.

Sohail used a different method, saying:

> 57, 8 to 65, 10 to 75, that's 18 altogether.

Brian described his method:

> I started at 75 and worked backwards, 5 to 70, 10 to 60, and 3 to 57, that's total 18.

Each member of the group then made subtractions for the others to work, using their number lines. Then I gave them '91 − 19' to work out in their heads, mentally. Jamil was the first to answer:

> I added 1 to each, that's 92 − 20, and that's 72.

I asked him if '91 − 19' had the same answer as '92 − 20'. 'Of course', he said. 'If you have 9p and I have 6p, the difference is 3p. If someone gives us each 1p, then you have 10p and I have 7p, and the difference is still 3p.' I told Jamil that he had invented a good method. The pupils now had three methods of subtraction they could use.

INSET

As mentioned before, adults often find mental subtraction difficult. Teachers are no exception! Careful study of the number tables is also important, so trying out any/all of the earlier activities is to be recommended. Certainly you should try mental subtraction in your INSET group, and think how the method of decomposition might be introduced. Identify as many different methods as possible and plan to encourage your pupils to use paper and pencil to investigate and describe the various methods they discover.

Multiplication and division

Some children memorize all the multiplication facts up to 10 × 10 before Level 4, and although most pupils will memorize many of their multiplication facts during the following year, it is convenient for us to discuss the process of learning these multiplication facts here.

Some children find it easy to learn their tables; for others this task is boring. So, it is encouraging to explain to them that there is a minimum which everyone should know.

(Some teachers will then allow the use of calculators – but will continue to give problems to help the children to learn.)

Obviously, it is important they should be confident in their knowledge of addition and subtraction facts up to 100 before they extend their knowledge of multiplication and division facts to this level. Also, we must make sure that pupils understand that multiplication is a quick way of adding equal sets – particularly useful when shopping, if we are buying more than one article of the same kind.

We assume that, by now, the pupils can double numbers up to 50. If doubling is followed by halving, some pupils may need a little help, because although they may find it easy to halve even multiples of 10 (20, 40, . . . , 100), odd multiples may prove to be more difficult. If asked to re-write the odd multiples (e.g. $30 = 20 + 10$) so that half of 30 becomes $10 + 5 = 15$, and so on, they often solve this problem for themselves.

By now, also, the pupils can multiply any number up to 20 by 10 immediately. Check that they know what the final zero means in, say, $13 \times 10 = 130$. (Answer: It tells us that there are no units, there are 13 tens.) Many pupils can count in 5s to 100 without knowing, for example, how many 5s there are in 35. If they know their 10 times table, e.g. $10 \times 7 = 70$, they often suggest halving 70 to obtain 5×7.

Pupils will know some isolated facts as well, e.g. the squares of numbers up to 10, because they have used unit squares many times to make and investigate the patterns of the sequence of squares (Fig. 2.9).

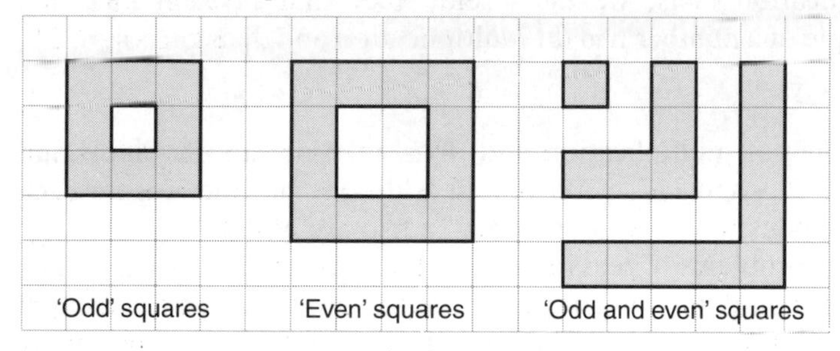

Figure 2.9

It is also useful to write the multiplication tables in different ways. For example:

1	**2**	**3**	**4**
0	0	0	0
1	2	3	4
2	4	6	8
3	6	9	12
4	8	12	16
⋮	⋮	⋮	⋮

and also the related groups

1	**2**	**4**	**8**
0	0	0	0
1	2	4	8
2	4	8	16
3	6	12	24
⋮	⋮	⋮	⋮

The table of 2s is found in column 2 by doubling; the 3s column is found by adding columns 1 and 2.

Pupils can be prompted to build these tables in different ways by questioning them. For example, when columns 1 and 2 are complete, ask them how they can construct the 3s column. 'Add the columns of 1 and 2' is the usual response. This done, ask them how they will make the 4s column; 'Double the 2s column', is usually the first response. Give them time to do this, then ask for another method of calculating the 4s. 'Add columns 1 and 3' is the next suggestion. (Having tried this, they often find doubling is easier, and will remember this for themselves.)

It helps the pupils to build the tables, if numbers are grouped whenever possible, e.g. 3, 6, 9, and the teacher takes time to discuss the relationships with them. Nine is an easy table to learn.

Activity 2.18 Ask the children to make a list of the multiples of 9 in order. They are then to add the digits and write each sum underneath each multiple. When they reach 99, the first digit sum is 18. But for this purpose we need a single digit, so they must add once more to reach 9. What can they discover about multiples of 9? How will this knowledge help them to memorize multiples of 9? (Many children mistakenly think that $6 \times 9 = 56$.)

When looking at multiplication facts, we must remember that **division facts** are important too. One example of a **number trio** for multiplication and division is 4, 6, 24:

$6 \times 4 = 24 \quad 4 \times 6 = 24 \quad 24 \div 4 = 6 \quad 24 \div 6 = 4$

Pupils can explore examples of multiplication and division trios, and addition and subtraction trios too, e.g. 2, 7, 9 and 2, 7, 5. Write four additions/subtractions for each trio, and consider what happens when two numbers of a trio are equal.

Children should be given problems of real interest to them as often as possible. For example, money problems should use current prices. Pupils can be asked to research the prices of videos, calculators, books, CDs, etc., and thus find out how many calculators, say at £4.25, can be bought for £20. Also, what change would they get?

Problems on the calendar can also be realistic, e.g. the Mason family go on a fortnight's holiday on August 10th. When do they return? The Scotts go one week later, also for a fortnight's holiday. What are the dates of their holiday?

Another useful task is to ask the pupils to make a multiplication square (Fig. 2.10) for the numbers 0 to 10, and then to look at the numbers on the major diagonal.

- Do they recognize these?
- Why do these numbers occur on this diagonal?
- What is the pattern of these numbers?

Members of an INSET group could equally be asked to make a multiplication square and to colour in the squares containing the number 12.

- Are these numbers in a straight line?
- If not, draw a curve through them.
- Does the curve have mirror symmetry?
- Put in the mirror line.
- Use a different colour to mark each occurrence of a 6.

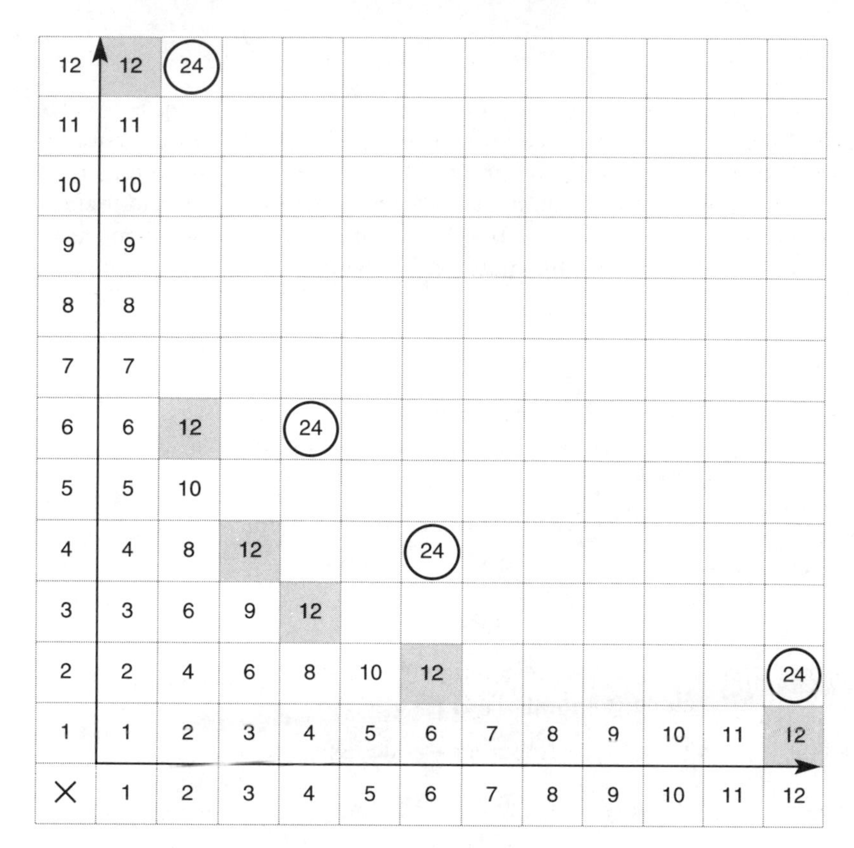

Figure 2.10 *Multiplication table*

Progression in written calculations

Most of this section is derived from the Non-statutory Guidance. It begins by emphasizing that in order to progress through the Levels, pupils at every stage should be encouraged to develop their own methods for carrying out calculations.

By Level 3, most pupils should recognize the types of problems they are dealing with and be equipped with the necessary number knowledge. As they grow in confidence and understanding they should refine and develop their methods, building up a range of ways of tackling calculations. Their number knowledge should begin to acquire structure and to be readily available.

It is important that they should be encouraged to make decisions about the best way to tackle each calculation. The methods are grouped as: mental; paper and pencil; and calculator methods. Methods adopted frequently contain elements from each, since, in practice, both children and adults employ a mixture of techniques when doing calculations. Such flexibility should be encouraged.

Addition

Usually pupils do not have difficulty with written additions, even when these involve 'carrying figures'. However, when first faced (at Level 4) with the addition of several single figures, many pupils resort to finger counting, 'to make sure'. These pupils need short daily oral practice until they can respond quickly and accurately. Remind them of quick addition of 10, of 9, of pairs of numbers with sum 10, of doubles, and near doubles. If necessary, make tens and units material available.

Subtraction

Case 2.3 These pupils (at Level 3) were provided with real money when developing the shopkeeper's method of giving change. Their teacher asked them to record what they had done. They recorded:

I bought a calculator for £4.45. I gave the shopkeeper £10. Find my change.

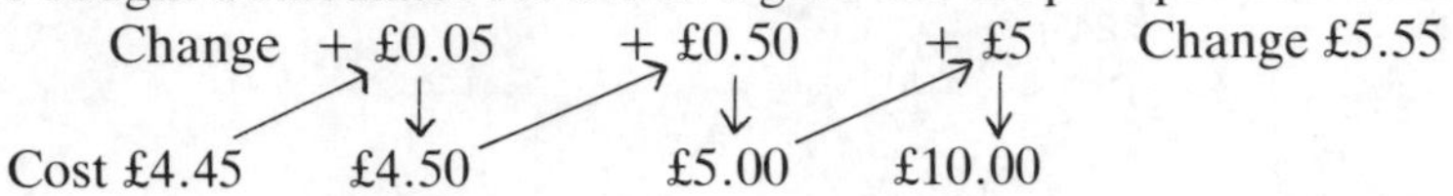

After the pupils had successfully recorded a number of examples, the teacher asked them to work 93 − 48 by a similar method. They recorded:

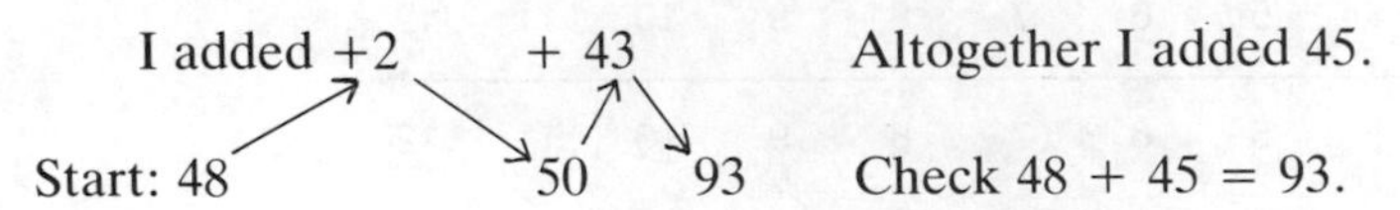

Gradually the teacher introduced harder examples: 382 − 178. Jean wrote:

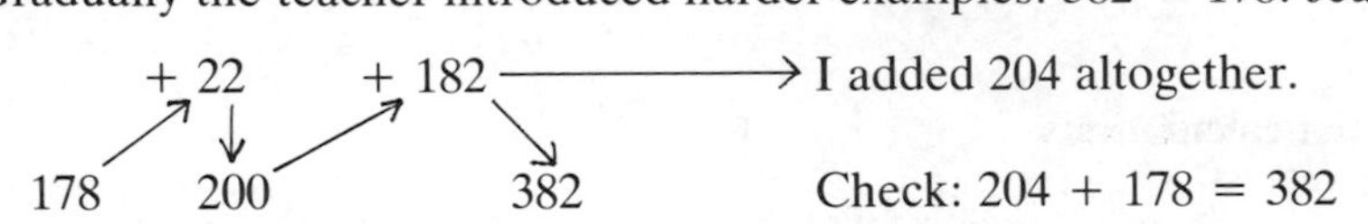

The teacher had given the pupils practice in rewriting a variety of numbers. For example, 83 = 80 + 3, or 70 + 13, or 60 + 23, and so on. She asked them to think whether this would help them to find the answer to 72 − 25, 91 − 19, 64 − 37, and so on. Suzanne recorded:

$$\begin{array}{rl} 72 \rightarrow & 60 + 12 \\ -\,25 \rightarrow & -(20 + 5) \\ & 40 + \;\;7 \rightarrow 47 \end{array}$$

decomposition method

$$\begin{array}{lr} \text{Jamil remembered: } 91 + 1 \rightarrow & 92 \\ \text{subtract } 19 + 1 \rightarrow & -20 \\ \text{difference} & 72 \end{array}$$

equal additions method

There was much discussion about the merits of the three different methods. However, when the teacher began to ask the pupils to subtract two-digit numbers mentally, most of them (except for Jamil) preferred the addition method. This was probably because they had originally relied on a number line to help them to devise a mental method (Level 4).

Multiplication methods

Case 2.4 A class of 10-year-olds was finding out whether or not their classroom floor contained more than 1000 square tiles. They counted 38 tiles in one side and 37 in the other. Jeremy and Rosemary began to count all the tiles but Toby persuaded them to draw a picture (Fig. 2.11) and to divide it into easy bits.

They then found the number of tiles in each section by multiplying. The teacher took the opportunity to show the class that the picture they had drawn was like 'long multiplication'.

38 × 37	30 + 8 ×			38 ×
	30 + 7			37
	900 + 240 →	1140	(30 + 8) × 30	1140
	210 + 56 →	266	(30 + 8) × 7	266
Total	900 + 450 + 56	1406	(30 + 8) × 37	1406

There are 1406 tiles – 406 tiles more than 1000.

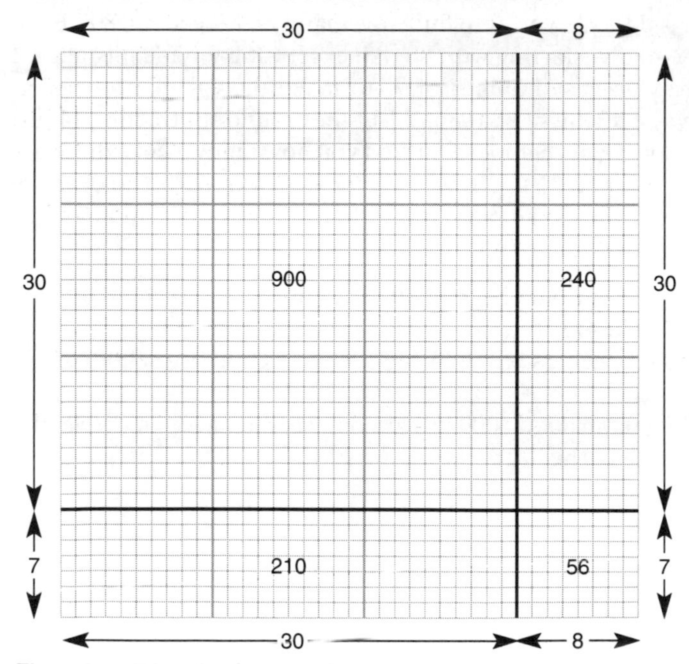

Figure 2.11 *Tiling the classroom floor*

Most pupils chose to multiply two-digit numbers (e.g. 57 × 69) by the 'number of tiles' method and a few applied it when multiplying a three-digit number by a two-digit number (Level 4). Once the use of calculators was encouraged, they changed their minds.

Division (Level 5)

There are two different problems which require division for their solution, each with an appropriate language pattern. The problems involve sharing, and repeated subtraction. When focusing attention on the sharing aspect of division, the pupils need to be provided with lengths of tape, sheets of paper, lumps of clay or Plasticine, sets of five identical glasses.

Activity 2.19 First ask the pupils to halve each quantity in turn, then to quarter it. Ask them to make a display; remind them to display the whole in each instance. Discuss remainders. Why are there none? Then ask them to halve different numbers of countable objects, e.g. beans or identical cubes. You may want to take this opportunity of introducing, or revising, the probability of taking an even (or odd) sample of cubes and of continuing with the probability of taking a multiple of 3, etc. Make sure that the pupils are confident about when to record remainders as numbers, e.g. 'One quarter of 77 is 19, 1 left' (Level 3). Ask the pupils how they would record 98 ÷ 4. Provide practice.

For the subtraction aspect of division a different approach is needed.

Case 2.5 I asked some slow 9- and 10-year-olds: 'It is 80 days to Christmas. How many weeks and days is this?' John said, 'We've got to find how many 7s there are in 80.' Each then began to subtract 7 repeatedly, starting with 80. All except John made mistakes. I asked if they could take away an easy number of 7s to make the calculation easier. 'Yes', said Pat, 'let's take away a fortnight at a time.' But they found subtracting 14 just as difficult and were soon discouraged. 'Suppose I asked you how many weeks there are in 70 days, would you know?' I asked. 'Yes, 10 weeks', they answered. 'So 80 days are 11 weeks, 3 days.' I gave them other examples (more than 70 days) and all began by subtracting 70. Then I gave them a range of problems.

- How many coaches (holding 42) would be needed for an outing for a school of 500?
- How many if the coaches held 47?
- How many boxes holding 12 crackers are required for every child of a school of 450 to have a cracker?
- A notice in a lift reads 'Maximum load 1000 kilograms'. How many men of 'weight' 78 kg could be carried safely in this lift?

Recording

Maximum 1000 kg

$$\begin{array}{r} 10 \times 78 = 780 \\ 2 \times 78 = \underline{156} \\ \underline{936} \end{array}$$

$$\begin{array}{rl} 1000 & - \; 936 \text{ kg} \\ \underline{-\;936} & \\ \underline{64} & \text{kg remainder} \\ 12 & \text{men} \end{array}$$

I then asked the pupils to make up their own problems. There was a good response. I also showed them the traditional method of long division. They all thought 'our' method easier. After this we used calculators. They hailed division with whole-number answers as very easy, but when I asked them to check our original problems they found the decimal 'remainders' very confusing. We postponed these!

FRACTIONS, DECIMALS AND PERCENTAGES

Have you ever wondered why there are three different ways of expressing non-whole numbers? Fractions were invented first; unit fractions, 1/2, 1/3, 1/4, etc., were known and used by the Egyptians as long ago as 1550 BC. They made up many interesting problems about fortunes left by wealthy men to their children! The ancient Greeks also worked with fractions – and the Romans, who based their taxes on fractions (1/50 on the sale of a slave, and 1/100 as a land tax) were close to inventing percentages. The word fraction means 'broken' because a fraction of a number or a quantity is not as much as one whole – it is only a part of a whole.

It is important to be able to change from fractions to decimals and percentages, and to know their relationships to each other. The complete name for a decimal is a decimal fraction because decimals, too, are only parts of a whole – but decimals are special fractions based on tenths, hundredths, etc. of a whole.

Percentages were invented almost a century before decimals, round about the time of Christopher Columbus. As measuring instruments became more precise, the fractions recorded became too cumbersome to work with. To simplify commercial calculations, percentages were invented, enabling clerks to work in whole numbers for much of the time.

These are considered in more detail towards the end of this chapter.

Decimals and fractions (Levels 4, 5 and 6)

Fractions are initially introduced at Level 2 in the Measures: finding 1/2 and 1/4 of lengths and areas by folding, halving masses by using balance scales, and halving yogurt pots of water by matching levels.

By Level 4 pupils should be accustomed to calculating with simple fractions: 1/3 of a pint of milk, 1/2 a metre of ribbon, 25 pence as 1/4 of £1. They should be able to add 1/2 and 1 1/4 metres of ribbon, and find the difference between 3 1/4 and 1 1/2 metres of ribbon.

Decimal calculations

At Levels 4 to 6 most calculations will be carried out by using a calculator. However, pupils must be able to record their working and to understand the processes. Addition and subtraction present no problems, provided that pupils remember to write numbers with their decimal points aligned. Then decimal numbers can be treated in the same way as whole numbers: they belong to our place-value system.

However, when we multiply or divide decimal numbers we have to take more care with the placement of the decimal point in the answer. Calculations which involve only one decimal number and one whole number are straightforward. For example, the cost of 3 writing pads at £1.35 each will be 3 × £1.35, i.e. £4.05. The cost of 8 calculators at £2.55 each is £20.40.

When two decimal numbers are involved we have to take more care. For example, the cost of picture glass depends on its area, so if we want glass for a picture 21.4 cm by 16.3 cm, we multiply 21.4 by 16.3. Using vulgar fractions we have

$$\frac{214}{10} \times \frac{163}{10} = \frac{34\,882}{100}$$

which we write as 348.82. We have the same number of decimal places in the answer (2 places) as we do in the two numbers taken together (one plus one). Let us try another example: 0.6 × 0.28. We might expect 3 decimal places in the product since we are multiplying tenths by hundredths, making thousandths. So

$$0.6 \times 0.28 = 0.168$$

If, instead, we have

$$0.6 \times 0.14 = 0.084$$

we have to 'insert a zero' after the decimal point to obtain the necessary three decimal places.

Now let us try division, How many lengths of 2-cm slatting 0.23 m long can be cut from 10.5 m? We have to calculate 10.5 ÷ 0.23. Our easiest method will be to divide by a whole number. We achieve this by multiplying 0.23 by 100, to give 23. But to obtain the same answer we must also multiply the numerator by 100. Then we have

$$(10.5 \times 100) \div (0.23 \times 100) = 1050 \div 23 = 45$$

Here are some more examples to refresh your memories.

Activity 2.20 Give answers to 3 sig.fig.

1. Add 1234, 1.234, 123.4, 0.1234 and 12.34. (1370) Subtract the smallest from 1. (0.877)
2. (i) 0.8 × 1.23 (0.984)
   (ii) 1.2 × 0.05 (0.060)
   (iii) 11.2 × 0.3 (3.36)
   (iv) 0.04 × 0.01 (0.0004)
3. (i) 25.6 ÷ 0.16 (160)
   (ii) 172.8 ÷ 12 (14.4)
   (iii) 0.96 ÷ 0.08 (12)

So far, children have been asked to use decimal fractions (usually to two decimal places) to write sums of money and also in measurement, or when using calculators.

When using decimal fractions to express sums of money, it is important to remember that the pound is the unit (unless the sum is less than £1, e.g. when we may write 60p). Therefore we do not write £1.60p (mixing the units) but £1.60. Children need to be reminded about this.

Our place value system can be extended beyond 1 to include fractions, shown here as (vulgar) fractions and decimal (fractions):

1000	100	10	1	1/10	1/100	1/1000
				0.1	0.01	0.001

Moving to the left we multiply successive terms by 10; to the right we divide successive terms by 10. As shown, in decimal form $1/10$ is written as 0.1, $1/100$ as 0.01, $1/1000$ as 0.001, and so on. To avoid confusion, we usually write a zero in the units place for numbers less than one, i.e. we do not write .1 or .01, but 0.1 and 0.01.

Activities 2.21 Ask the pupils in groups to cut five squares, each 1 decimetre square, from an A4 sheet of centimetre-squared paper. Ask them to colour 1 sq cm in the corner of one of their squares, and to calculate what fraction this is of the whole square. (They should give both the (vulgar) fraction ($1/100$) and the decimal equivalent (0.01)). They should label the coloured square clearly. Next ask them to take the same square and label 0.1 of that whole square, to colour that area and to label it as in Figure 2.12.

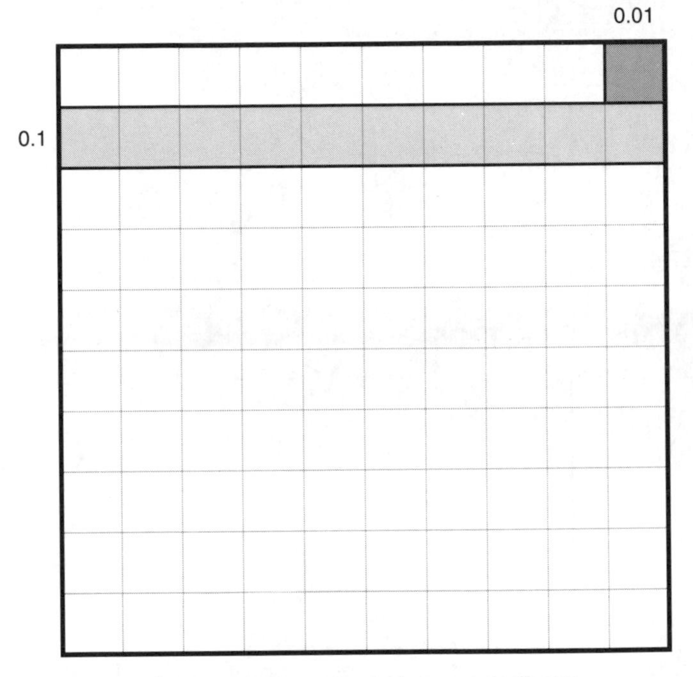

Figure 2.12 *Decimal fractions*

The pupils can then consider the total area shaded ($0.1 + 0.01 = 0.11$), and the difference between the shaded area and the whole square ($1.00 - 0.11 = 0.89$).

The second square can be used to form a pool of puzzles for the class. Ask each pupil to use one colour to make a picture on their second square and to make a note (not on the square) of the size of the areas coloured and left plain, and the sum of and difference between these areas. Remind the pupils to write their names on the back of their puzzle picture.

The third square can be used for multiplication; they make a pattern in one corner, repeat this in the other corners, then find the total area of the patterns by multiplication (Fig. 2.13).

The fourth square can be used to illustrate division, e.g. ask how many areas of 0.03 sq dm can be cut from 0.1 sq dm; from 1 sq dm (33, with 0.01 sq dm left). So, $1 \div 0.03 = 33$, remainder 0.01 (Fig. 2.14). Give frequent short practices.

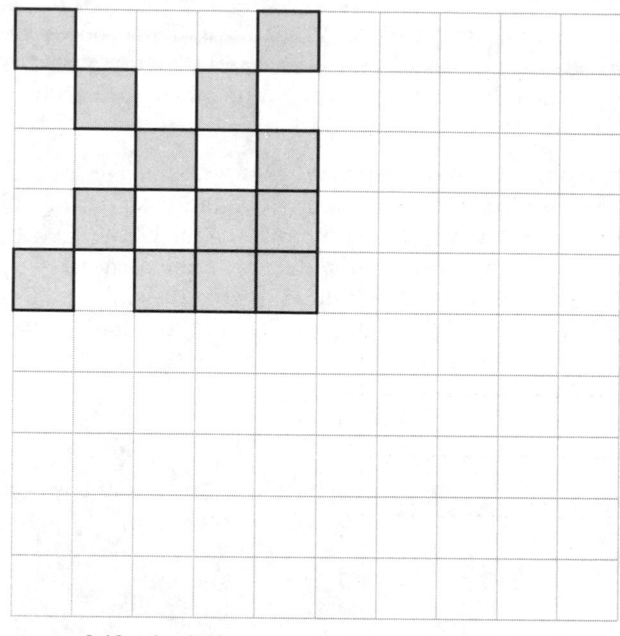

Figure 2.13 *Multiplication*

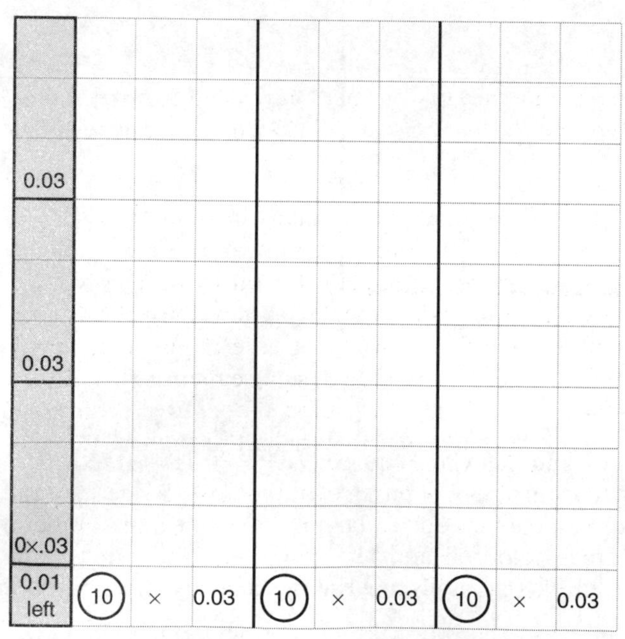

Figure 2.14 *Division*

The fifth square can be used to check pupils' understanding of decimal fractions. Ask them: which is greater, 0.6 or 0.28? Ask them to shade 0.6 in one colour and 0.28 in another colour, overlapping. Then ask them first to work out the difference between 0.6 and 0.28 as a calculation, and then to check this using their diagram (Fig. 2.15)

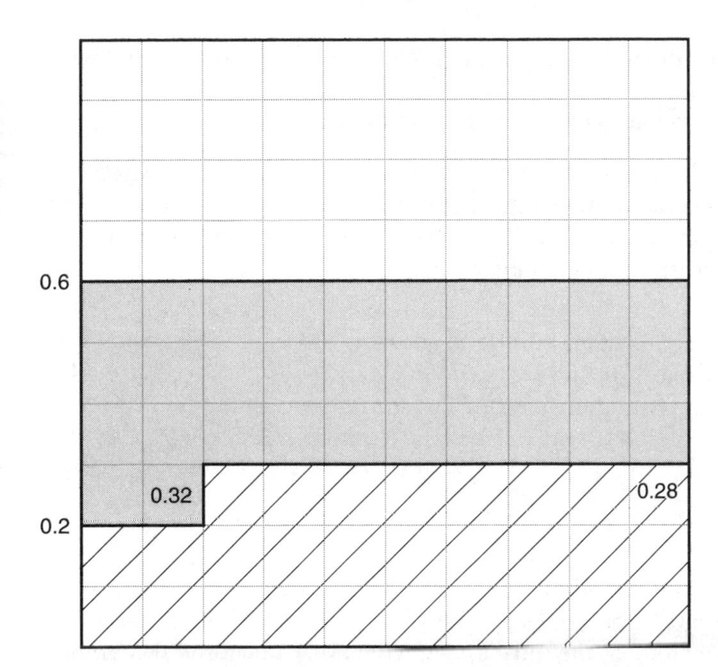

Figure 2.15 *Subtraction: which is greater, and by how much? 0.6 or 0.28? 0.6 − 0.28 = 0.32*

(Level 4). Some may need to use the diagram first. Others may need to write 0.6 as 0.60. Repeat problems of this type until all pupils are confident and can respond quickly and accurately.

The next activity involves a calculation of skin area, but perhaps should not be used if any pupil has attended a burns unit or has some personal experience of their work. Find out whether the local hospital has an approximation table for burns which they are willing to give you. The St John Ambulance Brigade's approximation is that the area of the sole of the foot is 1 per cent of the skin area. (Wait until the pupils understand percentages!) When the pupils have found their skin area in square metres (to 2 d.p.), ask the shortest and the tallest pupils to cut rectangles in newspaper equivalent in area to their own skin area, with one edge matching their own height. They can test the rectangles for good fit. Make calculators available.

Estimation and approximation

As an extension of place value (Level 4) the following targets may still need to be attained:

- understanding and using the effect of multiplying whole numbers by 10 and 100;
- understanding and using the relationship between place values in whole numbers;
- solving addition, subtraction, multiplication and division problems using numbers with no more than two decimal places.

For Level 5 the targets are:

- multiplying and dividing, mentally, single-digit multiples of powers of 10 with whole number answers;
- approximating, using significant figures or decimal places.

Some of the Programmes of Study in this section will be met 'on the way' when the pupils are engaged on another activity in this section. Others, such as the one just covered, will need regular practice for a period. For some pupils it is better to postpone the concept for a time and return to it again later on.

Activity 2.22 Practise writing equivalent numbers, according to their place value. For example, 7000 is 7 thousands, or 70 hundreds, or 700 tens, or 7000 ones. Finally, 2 000 000 is 2 million, or 2 thousand thousand, or 20 hundred thousand, and so on.

Approximation

Which decimal place do you consider when you want to give an answer correct to the second place of decimals? Find 1/7 on your calculator: 1/7 = 0.14285714. To give the answer correct to 2 places, you look at the third place, which is 2. Because this number is less than 5 the answer will be 0.14. To approximate to 3 decimal places, you look at the fourth decimal place, which is 8. This is greater than 5, so the answer is now 0.143. (We are saying that 0.14285714 is closer to 0.14 than 0.15, and closer to 0.143 than to 0.142. The following digit determines whether we round up or down.)

Activity 2.23 Turn the following fractions to decimals (unless you know them already!). Give answers correct to 3 d.p.

$\frac{1}{2}$ $\frac{1}{3}$ $\frac{1}{4}$ $\frac{1}{5}$ $\frac{1}{6}$ $\frac{1}{7}$ $\frac{1}{8}$ $\frac{1}{9}$ $\frac{1}{10}$

(Some of these answers are called recurring decimals. Can you see why?) Here is one worked example for you to check after you have tried it yourself: 1/6 (1 ÷ 6 on the calculator) equals 0.1666666. Taking 4 d.p., this becomes 0.1667; to 3 d.p. it becomes 0.167.

Significant figures

In experiments (e.g. statistical or scientific) we are often asked to express our answers to a given number of significant figures, e.g. 3 sig.fig.

The first significant figure is the first non-zero digit within a number. Leading and trailing zeros are needed to retain the correct magnitude of the number. Then, for 3 sig.figs only 3 figures other than 0 (unless 0 comes between the first and third figure) must be quoted, the first significant figure together with the next two (rounded if necessary). The middle figure, the second of 3, can be zero, e.g. 6018 is written as 6020 to 3 sig.fig.; 0.0064129 is 0.00641 to 3 sig.fig.

Activity 2.24

1. Write the following to 3 sig.fig.

   (i) 9.308 and 9 308 000
   (ii) 8143 and 0.0143
   (iii) 5 060 000 and 0.00506
   (iv) 2997 and 0.2997

2. Write 5 800 000 to 1 sig. fig.

Estimation

Estimating and approximating are needed to check the validity of addition and subtraction calculations (Key Stage 3) and to check that answers to multiplication and division problems involving whole numbers are of the right order.

Activity 2.25 In an INSET session, discuss the following approximations.

1. In a primary school there are 195 boys and 233 girls. Find the total and approximate to the nearest 100. Find the difference and approximate to the nearest 10 (Level 4).
2. Estimate 218×29. There are different degrees of approximation: $200 \times 30 = 6000$; or $220 \times 30 = 6600$. Which is nearer? How do you know?
3. Estimate that $358 \div 29$ is about 12 (Level 6).

Approximating in order to be able to give a good estimate of an answer is a valuable habit to encourage in your pupils. They will need plenty of regular practice at first. Encourage them to approximate to different degrees – crudely at first – then more accurately. Sometimes the children will use a calculator to check but it is always useful to make a rough check first.

Vulgar fractions

It is important to ask the pupils to talk about what they are doing and then to record in symbols, e.g.:

1. Half a glass of water and another half glass make a whole glass; they record:

   $\frac{1}{2} + \frac{1}{2} = 1$

2. The difference between a full glass and a half glass is half a glass:

   $1 - \frac{1}{2} = \frac{1}{2}$

3. How many half glasses are there in a whole glass?

   $1 \div \frac{1}{2} = 2$

4. Two half glasses make a full glass:

   $2 \times \frac{1}{2} = 1$

More difficult calculations with fractions could be recorded as follows.

Addition

$$5\tfrac{3}{4} + 2\tfrac{1}{2} + 1\tfrac{3}{8} = 5\tfrac{6}{8} + 2\tfrac{4}{8} + 1\tfrac{3}{8}$$
$$= 8\tfrac{13}{8}\ (\tfrac{8}{8} = 1)$$
$$= 9\tfrac{5}{8}$$

Subtraction

$$4\tfrac{1}{4} - 2\tfrac{3}{8} = 4\tfrac{2}{8} - 2\tfrac{3}{8}$$
$$= 3 + \tfrac{8}{8} + \tfrac{2}{8} - 2\tfrac{3}{8}$$
$$= 3\tfrac{7}{8} - 2$$
$$= 1\tfrac{7}{8}$$

Notice that we use equivalent fractions when adding or subtracting; we turn all fractions to the same family of fractions – in our examples, eighths. We also had to know that there are 8 eighths in one whole, and to know that $7/2$ is equivalent to $3\,1/2$, for example.

Check also that your pupils at Levels 4 to 6 can work out simple fractions of numbers, quantities and money. For example:

1. Find $\frac{3}{4}$ of £6, £9, £15.
2. Find $\frac{3}{10}$ of 180 m, $\frac{1}{5}$ of 2 m 35 cm; $\frac{1}{4}$ of 300 ml, $\frac{3}{10}$ of 450 g.

Ratios, fractions, decimals

Multiplication

$$1\tfrac{4}{5} \times 4\tfrac{1}{3} = \frac{\cancel{9}^{3}}{5} \times \frac{13}{\cancel{3}_{1}}$$
$$= \frac{39}{5}$$
$$= 7\tfrac{4}{5}$$

Division

How many lengths of $4\,3/4$ cm could be cut from a $25\,1/2$-cm length? An 8-year-old girl invented this method.

'We could work in quarters (cm).

$$\frac{25\frac{1}{2}}{4\frac{3}{4}} = \frac{102}{19} \qquad 5 \times 19 = 95$$

5 strips of $4\,3/4$ cm, 7 lengths of $1/4$ cm left. That's $1\,3/4$ cm.'

Ratios (Levels 4–6)

Discuss the following accounts within your INSET group and decide their relevance to your pupils.

I have often been told by teachers when I question them about the meaning of ratio: 'Ratio is the chapter in the textbook I never reached!'

Activity 3.5 (page 53) offers a good opportunity to introduce ratios. Once the children have made their various strips to represent their own measurements I would introduce the term 'ratio', telling them that the ratio waist/neck was nearly 2. I then asked them what would be the approximate ratio of neck/waist. They suggested that the ratio would be $1/2$. I asked for another ratio which was about 2/1. The children offered cubit/span and continued. 'So the ratio span/cubit will be $1/2$.'

I asked each pair to make one set of strips (each labelled with the name of the part of the body at one end and the owner's name at the other end) into an ordered graph. They were to make the second set of labelled strips into a full-scale three-dimensional model of its owner. There were some distances they had to find in order to attach the strips in the correct places (e.g. on the height strip). Some fastened the strips together with Sellotape; the rest I stapled together. The children stuffed their models with crumpled newspaper. They hung the models along the corridor, with feet on the ground. Those who had not made a skeleton of themselves asked to do so. When they had cut another set of strips I asked them to make a half-scale model of themselves. 'How will you do this?' I asked. 'Fold all the strips in half and cut them,' was the immediate answer.

Another opportunity for using ratios is in creating scale models, e.g. of the classroom. The children were allowed to choose their own scale; some chose 1:50, others chose 1:20. They included desks, tables and cupboards. One pair used a small scale of 1:100 and were disappointed with the result. They found out from a book how to enlarge a model and completed this for Open Day (Fig. 2.16). Can you find out how this was done?

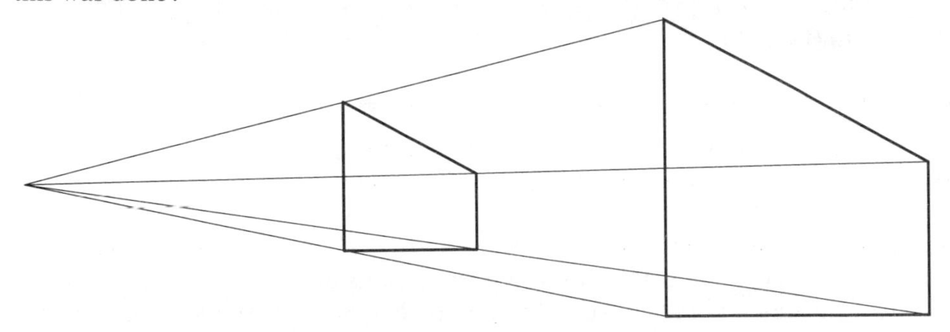

Figure 2.16 *Enlargement scale of playground (x2)*

Case 2.6 I asked groups of 6 pupils (Level 4/5) to make $1/10$ scale models of themselves. They measured the lengths they needed to the nearest centimetre, divided these by 10 and expressed the tenths in centimetres to 1 decimal place. They used cm/mm graph paper and cut the strips 1 cm wide.

When the models were complete, they decided to make the groups into pop, music, sports or drama groups – and made costumes and props to fit. When Lois looked at the size of her model's head, she said, 'Why is my model head so small? I could fit 100 of it on the top of my head!' This observation led them to investigate the effect on areas of reducing the lengths of the sides to $1/10$ scale. They drew a square dm, divided the sides by 10 and shaded the resulting corner square centimetre. The original area of 100 sq cm had been reduced to one (Fig. 2.17)!

They then turned their attention to the scale of maps in geography. Using maps of

Scale 1 to 10
Area: 100 squares reduced to 1

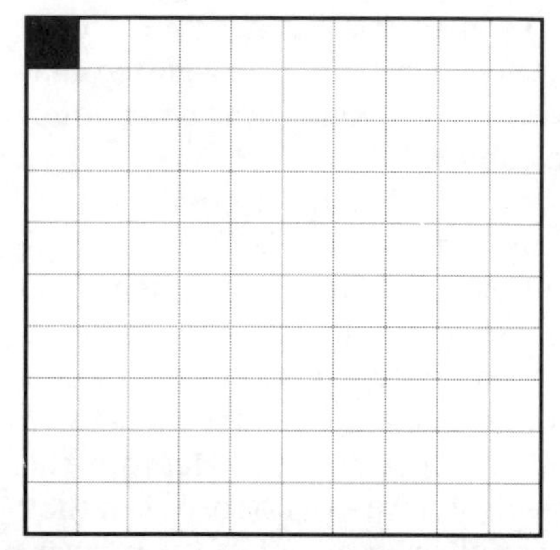

Figure 2.17

different scales, they confirmed that areas of villages, woods and lakes were reduced to the square root of the scale (which is always a fraction, e.g. 1 cm to 1 km, or 1 to 100 000, giving a ratio for area of 1 sq cm to 10 000 000 000 sq cm). Vocabulary: *inverses, squares and square roots:* $3^2 = 9$, $\sqrt{9} = 3$
cubes and cube roots: $2^3 = 8$, $\sqrt[3]{8} = 2$

Index notation (Level 5)

Powers of numbers are more often expressed as indices:

10×10 can be written as 10^2, and called '10 squared'.
$10 \times 10 \times 10$ can be written as 10^3, and called '10 cubed'.

Activity 2.26 How do you think that $10 \times 10 \times 10 \times 10$ is written, as a power of 10? (10^4) It is called '10 to the fourth', and succeeding powers are 10 to the fifth, sixth, etc., powers. Do you remember how to multiply powers of numbers? Work these to remind yourselves – then try division.

$x \times x^2$
$x \times x^3$
$x^2 \times x^3 \times x$
$x^2 \times x^4$
$x^3 \div x^2$
$x^4 \div x$
$x^4 \div x^2$
$x^6 \div x^2$

Decide yourself when your pupils are ready to be shown this method of expressing powers of numbers. Finally, write the map scale as a power of 10 (1 to 10^5); the area ratio is 1 to 10^{10}.

Percentages

There is no reason why decimals should not be introduced while pupils are still learning more about fractions, but percentages should be left until you think that the pupils are confident when working with fractions and decimals.

At Level 6 students are expected to be:

- working out fractional and percentage changes and finding one number as a percentage of another;
- expressing one number as a percentage of another;
- working out fractional and percentage changes.

Per cent means per hundred; so 50 per cent (written 50%) means 50 out of every 100, or 1/2.

We can turn fractions into decimals and then into percentages:

$\frac{8}{25} \rightarrow 32/100$ or $0.32 \rightarrow 32\%$

or percentages into decimals and then to fractions:

$76\% \rightarrow 0.76 \rightarrow 76/100 \rightarrow \frac{19}{25}$

We use the fact that 100% means 100 out of 100, or 1, to convert fractions straight into percentages – we multiply the fraction by 100%.

Activity 2.27

1. Write these fractions as percentages:

   $\frac{1}{4}$ $\frac{3}{4}$ $\frac{1}{8}$ $\frac{3}{8}$ $\frac{5}{8}$ $\frac{7}{8}$ $\frac{1}{40}$

   then use a calculator to write these fractions as percentages to 2 decimal places:

   $\frac{1}{6}$ $\frac{1}{3}$ $\frac{1}{9}$ $\frac{1}{15}$ $\frac{7}{9}$ $1\frac{1}{2}$

2. Convert these percentages to fractions (in their lowest terms, e.g. 52% = 13/25):

   75% $12\frac{1}{2}\%$ $67\frac{1}{2}\%$ $33\frac{1}{3}\%$ 16% 25% $2\frac{1}{2}\%$ 135% 180% 212%

3. VAT at 17½% is added to motor repair bills. Find the VAT on charges of £40, £60, £100. As a check, use another method. (Start with 10%?)
4. At a sale in a local store all goods are reduced by 40%. Calculate the sale price of goods originally marked: (a) £1.50, (b) £10, (c) £4.95, (d) £4.80. Calculate the reductions also.

Pupils will need short daily practice in converting fractions to decimal fractions and to percentages, and vice versa. They will also need practice in calculating fractional and percentage change, and in expressing one number or quantity as a percentage of another. As with all number topics, real-life situations should be used wherever possible.

Complete the following table:

fraction	*percentage*	*decimal*
–	66%	–
–		0.15
–	–	0.35
$\frac{3}{4}$	–	–
$\frac{2}{3}$	–	–
$\frac{1}{7}$	–	–
–	$7\frac{1}{2}\%$	–
–	–	0.85
–	$12\frac{1}{2}\%$	–

For example, it will be useful to introduce percentages during sale time in your locality. Local shops will have notices of coming sales, with discounts displayed. Ask the pupils to copy some notices. Once you have discussed with them the meaning of percentage, you can ask them to work in groups to make up problems based on the information they have collected. It would be useful to make a class book of problems.

Activity 2.28

1. A pullover marked at £7.50 is reduced to £5.50 in a sale. What is the percentage reduction on the original price?

   Reduction (as a fraction) $= \frac{£2}{£7.50} = \frac{4}{15}$
   (You may have written 200/750.)

   Reduction (as a percentage) $= \frac{4}{15} \times 100\% = \frac{4}{3} \times 20 = 27\%$ (approx.)

2. A school takes on three extra full-time teachers in September. There were 16 teachers last July. Find the percentage increase in staffing.

   Increase (as a fraction) $= \frac{3}{16}$
   Percentage increase $= \frac{3}{16} \times 100\% = \frac{75}{4}$ or $18\frac{3}{4}\%$

3. Labour and spare parts on Mr Wong's car come to £48. Find VAT at $17\frac{1}{2}\%$ (NB $17\frac{1}{2} = 10 + 5 + 2\frac{1}{2}$.) Also find Mr Wong's total bill.
4. Ask two pupils to copy the menu at a local restaurant, and ask them to find out the service charge. Ask them to choose a meal for their birthday (cost unlimited), and also a cheap meal. In each case ask them to display their choices with the costs and the total bill. Ask others to do this.
5. Mr Wood gives a discount of 10% for cash on bills of more than £10. Calculate the discount on bills of £12.50, £23.40, £37.50. What are the final bills?

Suggest that interested pupils should investigate Value Added Tax further.

- When was it first introduced, and at what rate?
- Which commodities were first taxed?
- Which commodities are not taxed in this way?
- When was the tax last increased?

Encourage the pupils to prepare some problems for the rest of the class to work.

Finally, here is the end of the Programme of Study for Number, Key Stage 2, published in the draft of the New National Curriculum, in December 1994. It is under the heading Solving Numerical Problems.

> Pupils should have opportunities to:
>
> a. develop their use of the four operations to solve problems, including those involving money and measures, using a calculator when appropriate;
> b. choose sequences of methods of computation appropriate to a problem, adapt them and apply them accurately;
> c. check results by different methods, including repeating the operations in a different order or using inverse operations; gain a sense of the size of a solution, and estimate and approximate solutions to problems.

This section deals with one important aspect of number which has often been neglected in the past – encouraging pupils to check their calculations and problems in various ways. Teachers should familiarize themselves with all the suggested methods before discussing these one by one with all pupils. An INSET session might produce some good ideas. A classroom display is also useful.

Pupils at Key Stage 2 will also be expected to test mathematical definitions and statements, e.g.:

- rectangles with the same area have the same perimeter (not true);
- it is harder to throw a '6' on a die than a '1' (not true);
- the product of three consecutive numbers is always divisible by 3 (true).

Not many teachers will have tackled problems of this kind, so they will need opportunities, as a group, for working activities of this type, and for making up such problems. At Key Stage 3 pupils are expected to make up card and board-games for themselves and to test the games out on others at the same level. They should also be able to collect and compare charts, diagrams and graphs from newspapers and magazines, to assess them and to be able to identify and rectify any misleading items. Teachers may therefore welcome a discussion group on topics of this kind.

Chapter 3

Measuring All Kinds of Objects

In this chapter, the stages of development are similar. This is an advantage, because to the children the activities seem refreshingly different while they serve to reinforce essential concepts already experienced.

The chapter includes all the measures (length, volume and capacity, 'weight' and mass, area, time and speed, and money). It is necessarily very long. However, you are not recommended to read it at one sitting. Rather, work through the first part which considers thoroughly all stages of development applied to length, and then consult subsequent parts for the other measures as you need them.

The Programmes of Study for the measures tend to be brief but we have dealt with them more extensively because we believe that teachers would prefer each part to be complete in itself. Moreover, problems based on the measures always have to be solved by practical methods and are therefore attractive to children, because they can understand how to reach solutions.

Some of the problems are exemplified by accounts of children from various schools solving similar problems. Accounts such as these often help teachers to formulate the stages of development and to become familiar with essential vocabulary.

INTRODUCTION TO MEASURE

In the December 1994 draft of the National Curriculum for Mathematics the Programmes of Study for the measures are included in the chapter on Shape and Space. The headings for each stage (Key Stages 1, 2 and 3) are identical. They are: Understanding and Using the Measures; choosing appropriate measuring instruments; reading scales to an increasing degree of accuracy; finding perimeters including the circle; finding areas and volumes by counting methods.

Stages of development

Examples here are taken mainly from length, but apply to all measures.

1. Understanding the concept
2. Conservation (of length)
3. Learning to recognize the different problems which can be solved by applying one or more of the four operations when comparing two objects, e.g.
   (a) by their difference (in length), if this is close;
   (b) by division (*ratio*) if the two are very different. Ask, 'How many of your feet tall are you?' Simple fractions: halves and quarters.
4. Arranging three objects in order (of length) (i.e. the need for a starting line)
5. Comparison (of lengths), using first non-standard, then standard, units; realizing the need for standard units of measure
6. Learning to read different types of scales
7. Appreciating that the smaller the measuring unit used, the more units are needed to complete the measurement – and vice versa (inverse relation)
8. Coming to terms with estimation and approximation
9. Using (a) fractions and (b) decimals in measurement
10. Finding the range, distribution and averages in measurement

KEY STAGE 1

During their first two years at school, children should have a variety of simple measuring experiences. Many of these will arise when they are comparing two lengths (heights, skipping ropes, perimeters such as neck lengths, etc.), two 'weights' on balance scales (shells, shoes, balls, etc.), two capacities (plastic cup or mug, bowls, etc.), two areas (scarves, tablecloths, books, etc.), two durations of time (time taken to change for PE, time taken for games, etc.). During these activities teachers will introduce the extensive vocabulary of comparison:

longer shorter taller
heavier lighter
holds more than less than
covers more than less than
takes more time less time

These comparisons are made in a variety of ways, all depending on direct matching.

LENGTH

We shall now work through the stages of development in measurement, using activities in length as the example. The latter part of the chapter looks at the other measures.

Understanding the concept (Level 1)

The concept of length as 'the distance between two points' is not always easy to accept. For example, it is confusing when we are thinking of the length of a perimeter fence. We are more aware of what the fence surrounds than of its length – unless we are paying for the fence! Teachers need to provide activities in which children have to find various perimeters (their neck and waist lengths, the perimeter of a running track or pond, and of a tree, etc.). These should be recorded as lengths.

Conservation of length

Young children often think that the length of an object depends on its position. For example, they think that they are taller when they are standing than when they are lying down.

Case 3.1 A teacher of 6-year-olds asked one of each pair of children to remove their shoes and to stand up straight against a clean notice board on the wall. The other one put a thin book on the first child's head, made a mark to show height and wrote the partner's name on the mark. They then exchanged roles so that all the children had a height mark.

Most children wanted to make a height strip. The teacher provided paper strips of different colours and the children cut and labelled height strips to match the 'standing height' of everyone in the class.

The teacher then asked the first group of children to return to their original positions against the wall and to lie on the floor with their feet against the wall where they had been standing. She asked their partners to check the length of each child. She watched the partners check each child's height against his 'standing height' strip. There was a surprised chorus of: 'They match. I'm still the same height!'

Comparison of two lengths

Case 3.2 (Level 3) We have already referred to children making and comparing height strips. The following record was written by a 9-year-old boy, using standard units of length.

> At 10 o'clock I measured the smallest boy's shadow – it was 57 in. His proper height is 50½ in. His shadow was 6½ in longer. At quarter past ten, his shadow was 54½ in. It had gone down 2½ in. I used the exact spot every time. I am measuring his shadow every quarter of an hour.

The results were accompanied by a carefully drawn block graph. The 9-year-old boy also wrote several questions for children to answer, using the block graph.

Introducing 'standard' units of measurement

As the children solve problems involving measurement by using differing measures of their own choice, teachers should begin to question them about their results.

Activity 3.1 Ask the children to measure the width of a narrow passage. If they decide to use their foot lengths, ask the child with the longest feet and the one with the shortest to carry out the activity. They should record their results on the same sheet, as in Figure 3.1.

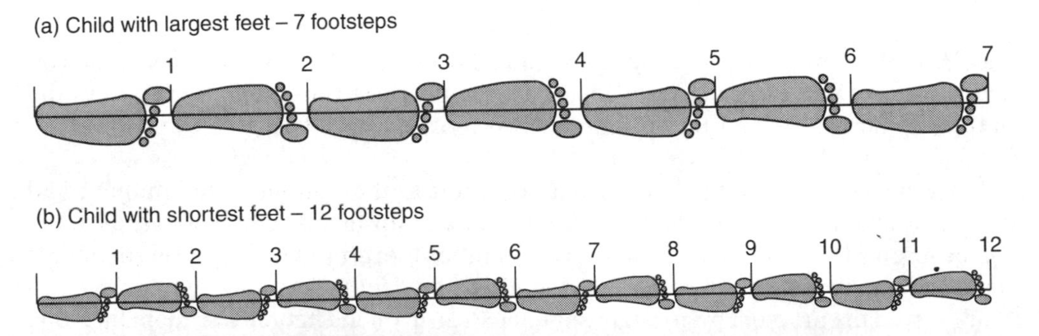

Figure 3.1 *Measuring the width of a narrow passage, by footsteps*

Suppose that the first child takes 7 foot-lengths and the second takes 12. Ask the children which result they should put on the record. Some children may suggest drawing round the longest foot, making seven copies in all and putting the seven footprints end-to-end to show the width! Others will realize that this would take too much time. Ask if any child knows a special (standard) length which everyone once agreed to use. Wait two days to see if any child responds – they usually do.

Problems solved by one or more of the four operations (Level 2)

Solution involving multiplication and addition

Case 3.3 James was 6½ years old. He had just measured the width and length of the hall – in feet because this was before we went metric. I asked James if he could now measure the height of the hall. He fetched a yardstick, stood on a chair waving the yardstick and saying: 'I knew I couldn't reach.'

The hall had been recently decorated with patterned wall paper and a dado, 4-feet high. I sat on the floor facing a wall and asked James to sit beside me. I told him that I knew he could not reach the ceiling but I thought there was another way of finding the height.

In less than two minutes James said: 'I've counted the patterns, there are 13. Now I'm going to measure them.' He soon came to tell me that the patterns were 9 inches long.

I wondered how a 6½-year-old would calculate the multiplication. I soon found out. He collected a surveyor's tape and chalk and asked me to come outside in the playground. He placed the tape in a long straight line, marked the starting point with chalk, then marked successive 9-inch lengths until he had reached the thirteenth. He read, '9 feet 9 inches. But I'll have to add the dado: 4 ft. That's 13 feet 9 inches altogether.' I congratulated James (and his teacher). This was a most exhilarating piece of work.

Solution involving division (ratio)

Another problem solved by 6-year-olds was: 'Find out how many of your own feet tall you are.'

Case 3.4 The children used various methods. Jason asked Lena to draw his outline on a long piece of newspaper as he lay on his back with arms by his side. He then drew the outline of both feet, and traced and cut two more pairs of feet, his estimate of his height.

Lena carefully placed the heel of one footprint at the bottom of the outline, and continued, heel to toe until she had placed the six footprints. She had not reached the top of Jason's head. She asked him to trace another footprint. He folded this in half. It just fitted in the gap. Jason recorded: 'I'm 6½ of my feet tall.'

Jill solved the problem by arranging her height strip on the floor in a straight line. She then put her heel at one end of her height strip and walked along the strip, placing heel to toe. She asked her partner Pam to check. She recorded: 'I'm more than 6 of my feet tall but fewer than 7.'

When every child had discovered that his height/foot ratio was between 6 and 7, the children asked their teacher if she would find how many of her feet tall she was. Her ratio was also between 6 and 7.

Jason asked the teacher if he could use her footprint to find out how many of her feet tall he was. She asked him to estimate how many of her longer footprint he would need. He replied: 'Your feet are longer so I shall need fewer. Say five.' Jason had made a good estimate – he used five tracings of his teacher's foot. The teacher realized that Jason had stumbled on another concept (7). She followed this up by asking the children first to estimate, then to check, the number of their foot-lengths they would take (heel to toe) to cross the corridor.

The need for a base line

When comparing two lengths (for example, two heights), it is important to introduce another concept: the need for a starting or base line.

Activity 3.2 When two children have compared their heights, using the right words, ask three children of different height to arrange themselves in height order, from tallest to shortest. When they have done this and described their positions (e.g. I am taller than Bob but shorter than Jean, etc.), ask them why it was easy to arrange themselves in height order. If they hesitate, ask the shortest to stand on a chair and then question them: 'Who is tallest now?' The answer is usually: 'That's not fair! You all have to stand on the floor.' They are beginning to understand the need for a base line. Give them other activities where a base line is needed.

For an assessment, ask the children to cut paper strips to match various body lengths, for example, perimeters of head, face, and waist. Then ask them to arrange their own strips in order from longest to shortest. Notice which children create a base line, using the edge of a table or a wall, etc. Some will just put the strips in a line by eye. The children will have made 'instant' graphs from their strips. Ask them to talk about these.

Standard units of measurement: metres, litres, kilograms, hours

In your INSET group, discuss the classroom activities on length which follow. If any of these activities are unfamiliar, try them within the group and discuss their value. Try them with the children you teach when an opportunity presents itself. Then within your INSET group discuss the purpose of each activity and the children's responses.

Materials needed It is an advantage to have a store of unmarked metre sticks – made with $2^1/2$ cm slatting. (You may already have encouraged the children to use these for 'measuring' distances before naming the metre length.) Also provide some strong paper strips a metre long, in case the children want to measure curved perimeters. These strips will also be useful when children need to measure half-metre lengths.

Activity 3.3 Begin by asking the children to try to find any object in the classroom which is a metre long and to record this. They should also record those objects which are slightly shorter and those which are slightly longer than a metre. Ask the children to make a display of their results. Suggest that they should include some curved metre-lengths.

Activity 3.4 Ask the children to make a careful estimate of the length and breadth of the classroom in metres. (They may suggest basing their estimates on the number of steps they take to cross the classroom.) Then show them the unmarked metre sticks. The first time they lay out the metre sticks across the classroom to check their estimate of the number needed to measure it, they will try to place them end to end in a straight line. If, at the end, there is a length shorter than a metre left uncovered, they usually neglect it the first time they try this activity. They record the two dimensions of the classroom in whole numbers of metres, and compare these with their estimates to discover whose estimates were nearest.

Before encouraging the children to measure to the nearest half-metre, you may want to check that they can find half a metre by folding a metre strip cut from paper. At the same time they may suggest marking the metre strip in quarter, half and three-quarter lengths. They will need some practice periods of estimating and checking lengths to the nearest quarter metre.

The next standard unit the children will need to use is the centimetre. The activity which follows makes use of measurement in centimetres. Each child in a group of four needs strips of a different colour of sugar paper. They will each need about 4 metres of strips 2.5 cm wide.

Activity 3.5 Explain to the children that once again they are to make coloured paper strips to match various body lengths. This time they will work in standard units of length. They should work in pairs, helping each other, and obtaining each length by matching. Ask them to cut one set of strips – perimeters of:

- head
- face
- neck
- waist (without pullover)
- ankle
- wrist

Remind them to write their own name at the end of each strip and the name of the part of the body at the other end. This set of strips can be used to make various comparisons.

1. Ask them to compare the perimeters of head, face and foot; and of waist and neck perimeters. Do they remember the approximate ratios?
   The strips can also be used to make a personal graph of the strips in order of length (without gluing them down). Ask each group of children to compare these graphs in as many ways as possible and to record likenesses and differences. Ask questions such as: 'Do all the graphs begin with the same part of the body? Do they all end with the same part?'
2. Ask the children to regroup the strips so that each pair of children in the group is responsible for the graphs of two parts of the body. Let them compare all the graphs. Finally encourage class discussions.
3. Ask them to arrange the strips for each part of the body separately in order of length, to record the lengths in order and to find the total length.
4. Ask them to begin with the longest pair of strips and to record their difference, and then to find and record all the other pairs of differences.

Different types of scale

Recognizing negative whole numbers in familiar contexts: a temperature scale, a number line, a calculator (Levels 3 and 4)

Measuring temperature

Celsius thermometers (room) should be available. Temperature scales are set out in columns in glass containers. The glass columns contain liquids which expand when the temperature rises and contract when it falls. The two normally used are mercury, a silvery liquid which is actually a metal, and alcohol, which is naturally colourless but is always coloured red for use in thermometers. In the Celsius scale the freezing point of water is 0° and the boiling point is 100°; in the Fahrenheit scale they are 32° and 212° respectively.

First ask the pupils to study the scale of a Celsius room thermometer. This scale is sure to range from a negative number (−20 to +50 perhaps). If the pupils ask about the negative sign, suggest that they count down from 10° to 0°, and then ask them what happens next. Some pupils suggest calling these the red numbers; others may tell you that these are minus numbers. Tell them that we call these numbers the *negative* numbers. Find out whether any pupil knows the name for the other (*positive*) numbers.

Give the pupils problems involving negative numbers based on thermometers. For example, ask: 'When the temperature rises from −5°C to 0°C, how many degrees is this?' (5°C) Ask if anyone can think of a method of showing that this represents a rise in temperature. Someone will probably suggest +5°C. Provide other examples and ask the pupils to set examples for each other, e.g. 'The temperature drops from +8°C to −4°C. How many degrees is this?' (−12°C)

Activity 3.6 Let the pupils work in pairs when using a room thermometer so that one can read and the other check the reading; then exchange roles. Make sure that they can tell you the interval between each of the gradations on the scale, and that they can approximate correctly to the nearest graduation.

1. This activity can be repeated in different parts of the school. First question the pupils about cold and hot places in their classroom, and in the school. Ask the pupils to try to find the hottest and coolest places in their own classroom. Remind them to record the place, time and temperature in each position. Suggest that they arrange their records in number order, from hottest to coldest, and that they record the difference in temperature between these.

   If the activity is repeated elsewhere in the school, first ask the pupils to record an estimate of the temperature before they begin to use the thermometer.
2. Suggest that each pair should find the average of their temperature recordings for the classroom. They can also find the mode (most frequent reading), and the median (middle one of a set of readings).
3. On a fine frosty day, ask the pupils to investigate temperatures outside; of water in different places (tap, puddle, water butt, etc.), and at different depths of the soil.

Return to measurement: distribution, range and central tendency

Introduce once again the measurement of each pupil's height and reach. Once more, the pupils can work in groups of four. Ask them what unit they will use and the degree of accuracy. Explain that when a younger class did this experiment, they divided the class into squares, tall rectangles and wide rectangles. Ask them if they can do the same. Ask them to find the pupil who is closest to a square in outline.

Name Height (cm) Reach (cm) Diff. (cm) Shape (square, tall or wide rectangle)

When all the records were complete, the groups were asked to make a table combining all the results. There was much discussion over this. Questions about scale were raised. Shall we measure the lengths to the nearest centimetre or half-centimetre? Suppose we have two lengths as close as half a centimetre? Would that count as a square?

Those who completed the combined table first were asked what other information they could add. Some found mean, mode and median for the combined results, and compared these three averages. Others cut full-scale height and reach strips and made a two-way graph with these. Height strips were placed vertically and reach strips horizontally. The graph took up so much room that it had to be made on the floor.

'Where do the squares come?' the teacher asked. 'Along the diagonal,' Joan said. The pupils then labelled the upper section for tall rectangles and the lower for the wide.

The teacher commented that through doing so many calculations involving one or two decimal places,the pupils realized that these are carried out exactly as they are with whole numbers only.

Directed numbers. The positive and negative number line

Positive and negative numbers are often called directed numbers when they are associated with a scale. On this scale zero is in the middle, positive numbers are on the right and negative numbers are on the left. Ask your pupils to help you make a number line on the floor of the classroom, using washable paint, as in Figure 3.2. How long should the steps be? 50 cm? Is there room for a line 10 metres long?

Number the scale from -10 to $+10$. Let the pupils give each other directions, in pairs. For example: Go to -3, then take $+4$ steps. Where are you now? They then record what they have done: $-3 + +4 = +1$.

To begin with, the signs denoting position on the number line are sometimes written above the number so that the position of the numbers in the positive or negative sections is clear, e.g. $\overset{+}{5}$, $\overset{-}{3}$.

Choose a position where the pupils can gather round to take part in the activities which follow. Let them help in the construction of the number line. Are they able to fit in steps from -10 to $+10$? Explain that they should use a positive sign (+) when moving forward in a positive direction → and a negative sign (−) when moving backwards in the negative direction ←.

Here are a few examples for pupils to try on the floor number lines. Let them make number lines that go from -10 to $+10$ in their books to use as an alternative. After an introduction let the pupils work in pairs, setting examples for each other. Vary these problems on thermometers/scales.

1. Go to $\overset{+}{5}$, then take $\overset{-}{3}$ steps. Where are you?
2. Go to $\overset{-}{4}$, then take $\overset{-}{2}$ steps. Where are you?
3. Go to $\overset{+}{5}$, then take $\overset{-}{2}$ steps. Where are you?
4. Go to $\overset{+}{6}$, then take $\overset{-}{6}$ steps. Where are you? At zero.

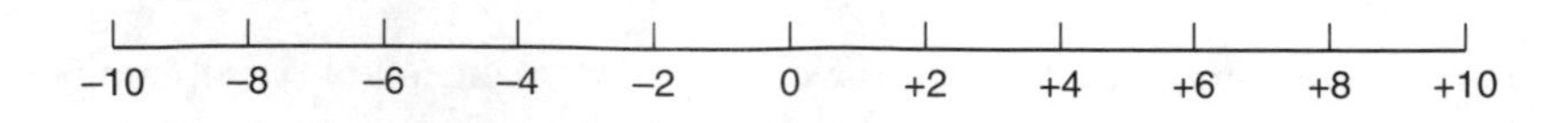

Figure 3.2 *Positive and negative numbers*

Now let us look at recording, for instance,

Start at $+4$. Go $+5$ steps. Where are you now?

Ask all the pupils to record this problem as

$(+4) + (+5) = +9$

Then

Start at -2, go $+3$ steps. Where are you now?

will be recorded as $(-2) + (+3) = +1$ and

Start at -10, go $+5$ steps. Where are you now?

will be recorded as $(-10) + (+5) = -5$

Explain that the children should use a positive sign when moving forward in the positive direction (+) and a negative sign when moving backwards in the negative direction (−).

Then ask the pupils, in pairs, to make up similar instructions for each other, and to record each activity. Sometimes they can give instructions in words, at other times in symbols.

The number line can also be used to demonstrate multiplication and division. It is best to start (or finish) at zero.

The instruction

Start at zero. Take 3 steps of -2. Where are you? (-6)

can be recorded as $3 \times (-2) = -6$.
For division, the instruction

Start at -10. Halve the distance to zero. Where are you? (-5)

is recorded as $(-10) \div (+2) = -5$ and the instruction

Start at $+8$. How many steps of -2 do you take to reach zero?

is recorded as $\frac{(+8)}{(-2)} = -4$ (4 steps backward)

Remember that your pupils will need regular short practice sessions with number lines for some time to come if they are to be confident in their use. Pupils can also make and use individual number lines from a metre strip of centimetre-squared paper 2 cm wide. The line can be labelled from -50 cm to $+50$ cm.

Reinforce these experiences by asking pupils to check their calculations, using a calculator.

Also, give further calculations to check. Pupils have probably come across negative numbers on calculators already, but suggest that they experiment with a calculator to find what they have to enter to obtain a negative answer on their calculator. Encourage them to record their successful attempts, and the final answers.

Different types of calculators have different systems; make sure that you are familiar with those your pupils use!

Remember: a positive sign (+) in front of a number does not change the sign of the number following; a negative sign (−) always changes the sign of the number following; check the examples which follow.

$$(-4) \times (-3) = (+12)$$
$$(-5) \times (+2) = (-10)$$
$$(+6) \times (-3) = (-18)$$
$$(+7) \times (+2) = (+14)$$

Check that the same rules apply for division.

$$\frac{-10}{-5} = +2 \quad \text{or} \quad \frac{-10}{-5} = \frac{(-5) \times (+2)}{(-5)} = (+2)$$

Smaller units of measure in the classroom

It is worth devoting an entire INSET session to small units, because it is essential that all teachers should have experience in reading and using instruments calibrated in small units. The co-ordinator should also make sure that teachers are familiar with a wide range of measuring instruments of different types and of different accuracies. They should be able to provide the right instruments for every problem they set their pupils.

Gradually, teachers will need to introduce their pupils to smaller units of length, of capacity, of 'weight', area and time. The pupils will soon be given the opportunity to choose the units they use, according to the problems they are set. Teachers will also provide problems which require an increasing degree of accuracy for their solution and introduce:

- measurement of length in millimetres;
- dial weighing scales (first in 10-g intervals to 1 kg, then in 5-g intervals);
- medicine bottles in 150-ml, 300-ml and 5-ml spoons;
- squared paper in square millimetres.

This process, which should involve many and varied activities, will occupy a substantial time, especially for some children. Problems such as the following could be solved.

Activity 3.7 Find 'Kwikfit' frames to take photographs and posters of different sizes. Does a catalogue exist?

Activity 3.8 Cut a square metre of paper from squared paper marked in square mm. How many square mm does it contain? How many (a) square decimetres; (b) square centimetres does it contain?

Activity 3.9 Investigate the relationship between a litre and millilitres; estimate the capacities of different medicine bottles. How many doses of 5 ml are there in each bottle? How long would each full bottle last when three doses are prescribed a day?

Activity 3.10 Obtain pamphlets from the Post Office with the most up-to-date first- and second-class postal rates for letters and packets. Make up a collection of letters and packets, and use postal scales to find how much each would cost to send by first and second class.

Estimation and approximation

Programmes of Study at Key Stage 2

You will notice that at Key Stage 2 the range of the Programmes of Study is considerably increased:

- making estimates based on familiar units;
- using a wider range of units of length, capacity, 'weight', and time;
- choosing and using appropriate units and instruments, interpreting numbers on a range of measuring instruments with appropriate accuracy;
- using decimal notation in recording money (e.g. £1.57 is correct, £1.57p is not).

Preliminary discussion should help pupils to realize that an estimate is not the same as a wild guess. An estimate is usually based on a careful comparison with a known personal measure, such as the length of a forefinger in centimetres, or of the palm of the hand for longer measures.

Activity 3.11 The first set of strips (used in Activity 3.5 above) can also be used for measuring in standard units. Ask the children first to estimate the length of all the strips in turn, in centimetres. Then provide metric tape measures or metre strips of centimetre paper and ask them to check their estimates.

Impress on them the importance of making careful estimates, rather than wild guesses. Let the pupils see that you take estimation seriously – pupils are normally interested in the outcome if they have first made a careful written estimate. Moreover, estimation helps to familiarize them with the unit under consideration.

Estimation makes pupils appreciate the importance of accurate measurement when using measuring instruments. Eventually, first teachers, then their pupils, should come to accept that all measurement has to be approximate, and that the degree of accuracy depends on the problems in hand. For instance, when painting a room, their parents will probably measure the walls in metres and take the next whole number of metres – to ensure that they buy sufficient paint. But when purchasing a bookcase which must fit into an alcove, they measure in centimetres, noting one fewer for the bookcase, to make sure that it fits into the space available.

Fractions

Your pupils have already worked with simple fractions. Also ensure that the pupils are learning to convert fractions to decimals. As usual, short daily practices over a period of time are important for sound learning.

Show the following number pattern to the pupils:

100 000 10 000 1000 100 10 1

Ask them to describe what is happening to the numbers as we move from right to left (we multiply by 10); ask them to write the next three terms to the left and to say these numbers aloud. Ask them if they recognize this sequence.

They will probably say: 'It's the place value sequence.' Next ask the pupils what the pattern is as we move from left to right. Now we are dividing each successive term as we move to the right. Are they able to continue successive terms to the right? Here is a larger part of the sequence:

1 000 000		a million
100 000		one hundred thousand
10 000		ten thousand
1000		one thousand
100		one hundred
10		ten
1		one
1/10	0.1	one-tenth
1/100	0.01	one-hundredth
1/1000	0.001	one-thousandth

Make sure that each pupil understands and can reproduce and label this place-value diagram correctly. Question them orally about the diagram at frequent intervals.

Range and averages

Introduce measurement of height (H) and reach (W) (finger tip to finger tip, arms stretched, with backs to a wall) in centimetres, then as a decimal fraction of a metre. Pupils can work in groups of four to find out whether their shape is a square (H = W), a tall rectangle $H > W$, or a wide rectangle $W > H$.

They can also find the average height for the group. If they cut strips to match their height, the groups can be asked to find the longest combined strip of 4. Then each group can be asked to find the average height, i.e. the height each group member would be if they were all the same height. Notice whether they folded the long strip in four to obtain the average. Ask each group to find out whose height is nearest to the average. List the heights of all the class in order. Repeat with the other measures.

Provide problems which cover all four operations (all lengths to be in decimal fractions of a metre) and a variety of the measurements they have made.

1. Find the average reach for the group. Record the differences between the height and reach for each group member.
2. Ask the pupils to double their neck perimeter and compare it with their waist perimeter, then to halve their waist perimeter and compare this length with their

neck perimeter. Repeat these activities with span and cubit. Display the group results in order.

Pupils should soon come to realize that calculations involving one or two decimal places, resulting from practical activities in the measures, are carried out in exactly the same manner as calculations involving whole numbers only. Provide regular short practice.

Finally, use the height strips to make a frequency graph of the grouped heights and the reach strips to make a frequency graph of the grouped reaches (3-cm groups?). Record the ranges of heights and reaches.

CAPACITY AND VOLUME

Capacity

Ask the children to help you to make a collection of a wide variety of clear plastic containers, to be used in the early stages of finding out about capacity.

Case 3.5 In this account 6-year-old children are comparing the capacities of a pair of plastic pots. I had asked them to find the difference in capacity as an amount of water.

Robert and Peter filled the larger pot with water and then poured the water into the (smaller) yogurt pot until it was full. Peter showed me the water left in the larger pot, saying, 'That's the difference in capacity.'

Mary and Helen filled the smaller pot with water, then carefully poured the contents into the larger jar. When I asked the girls how much larger the jar was than the smaller pot, Helen pointed to the air space in the jar. I agreed but reminded her that I had asked for the difference *in water*. Helen fetched a glass of water and a spoon and asked Mary to count the spoonfuls as she spooned water into the jar. Mary counted 11 spoonfuls to fill the jar and Helen ladled 11 spoonfuls into an empty glass which she labelled: Difference between the capacities of our pots, 11 spoons.

Roger and Jill filled both containers with water. They then looked for two identical jam-jars. All they could find were two fish-tanks. When each had poured the water in the pots into a fish-tank they were disappointed to find that the water was too shallow to see which contained more water. At last we found some identical jam-jars and they carried out the experiment again, this time using two identical jam-jars. Jill then spooned water from the jar containing more water into an empty pot until Roger told her the height of the water was the same in both jars. This pot was also labelled: Difference in size of these two pots.

I was delighted that these three pairs of children each devised a different method of solving the problem, particularly when the teacher explained that this was the first time these children had experimented with capacity.

Standard units of capacity

Later on, when teachers are about to introduce standard measures of capacity, it is useful to question the children about these units. They will probably know them from bottled drinks and can be asked to bring one example to school. We hope that the children will collect a wide variety of clear plastic containers: bottles of different shapes, bowls, jugs and vases. Let the pupils, in pairs, use a commercial litre measure to mark the litre graduation on all the litre containers.

Activity 3.12 Collect some pairs of identical litre containers. Ask the pupils to mark the litre containers first in halves, then in quarter and three-quarter calibrations. Ask them to describe their methods, and to use their marked containers to find others with capacities of 1 litre, 500 ml, 250 ml and 750 ml.

Volume

Problems involving volume may well arise while the pupils are solving problems involving capacity. However, some teachers will not have carried out practical activities involving volume; it is important that they should do so, before using them in the classroom. Begin by asking them to distinguish between capacity and volume. They need to understand the difference between these two concepts, and that their standard units of measurement are different: millilitres and litres for capacity, and cubic centimetres and cubic decimetres for volume.

Activity 3.13 Provide containers made of thick material, e.g. a vase or bottle made of thick glass or pottery. Ask the teachers to estimate the capacity of the vase in millilitres, and then the volume of the material of which it is made in cubic centimetres. They check by experiment. Ask them to discuss their methods.

Case 3.6 This account details experiments carried out by 8/9-year-olds on volume, and should suggest problems that can be investigated with your pupils.

Every pupil had brought a fist-sized rock to school to find its volume. In the playground arguments broke out between pairs of pupils about whose rock was larger.

John and Chris began by using string. Each was trying to find the longest perimeter of his rock. This took time because the rocks had awkward shapes. When they finally compared the longest perimeters, they found that these were almost equal in length. The teacher waited to see what the boys would do next.

Chris fetched two pieces of tissue paper and red and blue chalk. He gave the blue chalk and tissue paper to John and began to cover his own rock all over with red chalk. Then he wrapped his rock in tissue paper, carefully smoothing out the creases. John did the same with his rock. Both boys unwrapped the paper, smoothed it out, then cut around the edges of the chalk perimeters. They placed the red paper on top of the blue paper. A long piece of red overlapped the blue paper, but there was a piece of blue paper uncovered. When John cut off the long red piece and fitted it over the blue uncovered paper, the boys agreed that the surface areas of the rocks were almost the same.

But the boys were not convinced that the rocks were the same size. When the teacher asked them which rock was larger, they said that although the perimeters and the surface areas were nearly the same they did not believe that the rocks were the same size. The teacher then asked them what they meant by size and how else they could measure the rocks. John said that size meant the room rocks took up. He jumped up and ran out of the classroom. He soon returned with two metal files. He gave one to Chris and fetched an identical clean sheet of paper for each of them. They put their rocks on the paper and began to file them. They soon found that filing rocks was a very slow process.

The rocks were hard, especially John's, which was made of granite. The teacher now began to question the boys about their method. 'What will you do when you've finished filing?' she asked. John replied at once. 'You get two identical jam-jars, one for Chris and one for me. Chris puts his filings into his jam-jar and I pour mine into my jam-jar. We shake the filings until their surfaces are horizontal, then compare the levels.' The teacher was delighted with the boy's ingenuity but said, 'It will take a long time!'

She then brought them a clear plastic container half full of water and asked them if this would help them to find the size of the rocks. The boys hesitated, so the teacher asked them what would happen to the water when they lowered a rock into the container. 'The water will rise', they chorused. 'How much does it rise?' the teacher asked. Chris answered, 'The risen water is what the rock would be if it were made of water.' The teacher asked them how they could find and compare the 'risen water' for each rock. John said, 'We need two plastic bags. We stand the container in one bag, then fill it to the top with water, and lower a rock into the container. The risen water overflows. Remove the container and the rock and fasten the bag (with the risen water from my rock). Then Chris does the same with his rock.' When the boys had collected the risen water from their rocks, the teacher asked them how they could compare the volumes of the rocks, which both boys agreed would be the same as the volumes of the risen water.

First each boy tried to pummel the bag of water into the same shape and size as his rock. Then they poured the contents of each bag into two identical jars and compared the levels. Chris's rock was slightly larger.

Finally the teacher asked: 'Will the rocks each weigh the same as their risen water?' 'Of course,' they answered, 'because the rock has the same volume as the risen water.' The boys were sent off to check the weights of their rocks and of the risen water. Both returned puzzled. 'We must be losing water somewhere because the rocks do not weigh the same as the risen water,' they said. 'Please check our experiment.'

The teacher and a group of children checked that Chris and John had carried out their experiment carefully and had not lost any water. The teacher suggested that they should record the weight of the larger rock and of its risen water. Here are the results, in pounds and ounces, which were the weights they had.

Wt. of rock = 3 lb 12 oz
Wt. of risen water = 1 lb 8 oz

The teacher asked them: 'How many times as heavy as water is the rock?' John answered slowly. 'If the rock was twice as heavy as the risen water, that's 3 lb – too little. If the rock is three times as heavy as water that's $4\frac{1}{2}$ lb. That's too much. My rock is more than twice as heavy as the risen water – which is the same volume as my rock.'

This work, involving density, is defined as Level 6. However, the boys coped admirably because the problem was posed in a practical situation. Teachers will probably want to introduce volume by giving the children easier, more straightforward activities, such as the following examples. These should be tried out in an INSET session beforehand.

Activity 3.14

1. Ask pairs of teachers to bring to the next session two rocks of different shapes which they think have the same volume. Ask them to check.
2. Find the volume of a cricket ball. What would the 'risen water' from a cricket ball weigh if a litre of water weighs a kilogram?
3. Take 24 centimetre cubes and in your group make as many different cuboids of this volume as possible. Make a note of their dimensions. Also record the surface area of these cuboids.

Case 3.7 This next investigation was carried out by a class of 10- and 11-year-olds (Level 5).

The pupils noticed that their pet mouse seemed to spend far more time on feeding than the guinea pig did. Jerry asked if this might happen because the mouse was so much smaller than the guinea pig. He suggested investigating the surface areas and the volumes of a mouse and of a man.

But the teacher suggested that they should first find out how the surface area and volume of a cube changed as the cube grew larger. They began with a sequence of cubes having edges 1 to 10 units long. They worked out the number of squares in the surface area of each cube and related the surface area to the volume, and recorded them in a table (Table 3.1).

Table 3.1 *Comparing surface area and volume*

Edge length (cm)	0	1	2	3	4	5	6	. . .	10 cm
Volume of cube (cm^3)	0	1	8	27	64	125	216	. . .	1000 cm^3
Surface area (cm^2)	0	6	24	54	96	150	216	. . .	600 cm^2
S. area/vol. (sq. per cube)		6	3	2	$1\frac{1}{2}$	1.2	1	. . .	0.6

The pupils were excited by their results. They wondered whether the number of squares of surface area per cube would continue to decrease. They worked this out for a cube of edge 10 cm, and found the rate to be 0.6. They realized, from the table, that the surface area/volume rate for small creatures, like mice, was much greater than for a man. Small creatures must have to eat what seems to us a disproportionate amount of food to replace their heat loss.

Next term, these children returned to the 'mouse and man' problem. One father and his daughter Alison, aged 10, weighed everything they ate and drank for 24 hours. Alison's consumption of food and drink, considered against her mass, was considerably greater than that of the father measured against his mass. The previous term's findings had led the children to expect a result of this kind.

Activity 3.15

1. Pour a measured litre of water into a light plastic bag and fasten the top with a light clip. Use balance scales to find the mass of the bag and its contents.

   When the bag and clip are dry find their mass – or you can add an identical plastic bag and a clip to the other scale pan.
2. Find the volume of a kilogram of water.

How accurate were your results?

Case 3.8 Finding your own volume (Level 5)
A teacher of 11- and 12-year-olds asked them how they could find their own volume. They looked puzzled and Nina asked: 'Do you mean us to find the size of our skin?' Before the teacher could answer William began: 'Fill the bath to the top, then step in . . . ' He was interrupted by the class saying: 'William, your mother wouldn't let you!' So he began again. 'Put some water in the bath.' 'How much?' Nina asked. 'Four inches,' William answered. 'That's not enough to cover you,' Nina said. William continued: '12 centimetres then. I can ask my mum to check that I'm covered. If so, I can then ask her to mark the water level. Step out of the bath. Mark the new water level. This should be the same as the first. You have to find the volume between the marks.' The teacher asked William how he would do this. 'Fill empty litre bottles with water,' he suggested. Next morning, when William came into the classroom he announced, 'I've found my volume. I'm 38 litres.'

That morning, the teacher gave William a hollow clear plastic cubic decimetre and asked him to estimate its capacity in millilitres. He thought that it could not be as large as a thousand and was surprised when he found that the cubic decimetre held exactly a litre of water.

The pupils then decided to make 38 decimetre cubes in strong squared paper in order to make up a body shape of William with them. They made the life-size model of William's height. William made a label. It said: 'This is William. His volume is 38 cubic decimetres or 38 litres.'

MASS AND WEIGHING

While the following does not yet appear in the National Curriculum, we think that some teachers will be interested.

Units of mass: kilograms and grams

Mass is the amount of material an object contains and is measured in kilograms. The mass of an object depends on the material of which it is made and therefore never changes. Before introducing the kilogram, begin to talk about objects having more mass or less mass or the same mass as – instead of being heavier or lighter. Some materials (such as polystyrene) are very 'light'; others (such as lead) are 'heavy'. But of course a larger object has more mass than a smaller one made of the same material. Weight is a force caused by the pull of gravity towards the centre of the earth (or any other planet or moon). If we went in a capsule to the moon our mass would be the same there, but our weight changes as we climb higher up a mountain, or descend into a coalmine – or stand on the moon (or any other planet or moon). The pull of gravity is weaker up a mountain because we are farther from the centre of the earth. The unit of force is a newton. When you hold an eating apple (100 grams) in your hand, you are experiencing the pressure of 1 newton.

Primary children will not need to know the unit of force. However, it is important for them to know that kilograms and grams are units of mass. At the secondary school stage mass will be defined in terms of acceleration.

Try these sample activities in your INSET group, and discuss their value. They will need to be augmented and perhaps rearranged to suit your pupils.

Materials

Sturdy, top-pan balance scales with large pans are best for children to use. Many of these are too sensitive for children to balance one small object successfully against another. Provide objects with more mass for young children to use.

(In the early stages, when children are experimenting with balance scales on their own, find out whether each child knows which pan is holding the object with more mass. At first, many young children think that the higher pan holds the object with more mass.)

Activity 3.16 (Key Stage 1) The children can use their own possessions for this activity.

1. Ask them to find objects (a) with more mass than, (b) with less mass than, and (c) with the same mass as a shoe or other object of their own. Let them display their objects in three labelled sets.
2. Ask the children to make an object from Plasticine which balances a pencil-case.
3. Give each child a ball of Plasticine (fist-sized), and ask them to use balance scales to halve the ball.

   Notice whether, when the child has one piece of Plasticine in each pan, he takes pieces from the piece with more mass and adds to the piece with less mass until the two pans balance. Make sure that he does not forget the pieces he removed! He should leave the halves in the pan for you to check.
4. How many identical cubes balance 10 identical marbles? 20 marbles? Which have more mass, marbles or cubes? How do you know?

Activity 3.17 (Key Stage 2) Take two identical empty yogurt pots. Fill one with sugar and the other with white flour. Shake them down and refill to the brim. Is it true that sugar has twice as much mass as flour (old adage)?

Activity 3.18

1. Find a rock about the same size as two clenched fists. Collect water in a plastic bag which has the same volume as the rock. How did you do this? Compare the mass of the rock with the mass of the water in the plastic bag.
2. Ask your pupils to make a kilogram of Plasticine and to collect a kilogram of water in a plastic bag. Ask them to compare each of these with a metal kilogram mass. Which has the greatest/least volume?

   Does the mass made of the material with the least volume have the greatest 'density'?
3. Ask the pupils to collect empty kilogram packets of dry goods. What can you find out about the goods the packets contained?
4. Use a variety of spring scales, including bathroom scales, to find the mass of a newspaper. Discuss your findings.

Activity 3.19 (Level 6) Investigate the mass of a sheet of typing paper. Use 10, 20, . . ., 50 sheets. Can you think of another method? Now cut what you estimate to be one gram of typing paper. How many sheets would you need to make 10 grams? Investigate the mass of other types of paper. Which has the greatest/least mass?

AREA

Discuss in your INSET group what you mean by area. Did you include three-dimensional objects? What do you think of this statement:

> Area is the extent (or amount or measure of) a surface – in three or two dimensions?

Then discuss the early experiences of area which you plan for some of the school's youngest children. These should involve covering:

- a doll with clothes
- a bed with bedclothes
- a table with a tablecloth
- a picture with paint or crayon

Sample questions to ask are:

- Does the dress fit?
- Is it too tight?
- Too loose?
- Does the sheet fit the bed?
- Is it wide enough?
- Does the tablecloth cover the table completely?
- Have you painted the picture right to the edge?

It is important that pupils should have experience of covering surfaces of irregular shapes as well as regular ones. In this way they will come across problems which they will set about solving for themselves. For example, when covering a surface with square units, they will fold the whole units into fractions to fit the small spaces around the edges.

The following activities are planned for classrooms. Those teachers who are not familiar with some of the activities will need to work through them and discuss them within their INSET group. The vocabulary needed is:

area cover overlap larger smaller

Activity 3.20 Direct comparison Ask the children, in pairs, to try to find two leaves of different shapes which they think have the same area. Provide them with scissors and ask them to check. If leaves are not available, ask them which is larger, their handprint or their footprint. Use magazine paper for feet, plain paper for hands.

Activity 3.21 Comparison using identical non-standard units Ask the children, in groups of four, to find the child with the smallest footprint. Provide a choice of plastic shapes for covering surfaces: identical squares, circular counters, isoceles right-angled triangles. Children usually choose counters first. Observe how they begin covering their footprint. Do they fill in the edge first? When they stop, ask if they have covered all their footprint. (This may cause them to add a second layer!)

Offer them a second choice of identical shapes for covering the area. Their second choice is usually squares. This time they often fill the middle first, shifting the squares to minimize the awkward part-squares round the edge. Some children neglect these at first!

Standard units of area

The children will have found different answers for their areas so far, partly because they have not been using standard units, but also because in the first activity they chose different units. Now introduce them to standard units: a square-decimetre tile – the former 4-inch tile – is a convenient size. Kitchen and bathroom tiles are often square decimetres and these can be used as templates when cutting squares from magazines or newspapers.

Ask the children, in pairs, to work with a square-decimetre tile, and a copy cut from centimetre-squared paper. First ask them how close to a square decimetre in area the palm of their hand is. The palm of a young child's hand is about half a square decimetre in area. Let them compare the area of your palm with a square decimetre. Then ask them to try to find something else which is about a square decimetre in area. They can then find, from their square-centimetre paper, how many square centimetres there are in a square decimetre.

Activity 3.22 Ask each group to draw the outline of the back view of the smallest child in each group, using crayons.

Two members of the group should be cutting out square decimetres (100 sq cm), although there is no need to tell the children the name yet. Each group will need about 40 squares and each member should take a turn in cutting.

Ask the children to estimate the number of squares needed to cover the back view of the shortest child in their group. Observe how they made their estimate. Did they arrange a line of squares to cover the height? And perhaps another at the waist to complete the estimate? Discuss the methods used. This time the part-squares round the edges will be too large to neglect!

Now the children check their estimates. When they have filled in the whole squares, they often move them around to make sure that they have covered as much 'ground' as possible. Then without any direction they begin to fold whole squares into halves and quarters to cover the part-squares. They fold the squares along the diagonal or by matching opposite edges, according to the shape left uncovered. Then they pair the halves and put the quarters in fours before making their count. Ask them how they kept count of the total number of squares. Did they pile them in tens?

Activity 3.23 Finding a quick method of calculating the area of a rectangle. Ask the pupils to cut the following rectangles from centimetre squared paper: 1 cm by 2 cm, 2 cm by 3 cm, 3 cm by 4 cm, and so on. Ask them to find the area of each rectangle, and to record the areas within each rectangle. After they have found the areas of five or six rectangles, ask if anyone has found a quick way of calculating such areas. If they explain correctly (multiply the two lengths), suggest that they continue to use the method. Also ask two pupils to record the areas, in order, on the blackboard so that everyone may investigate the pattern.

Activity 3.24 To introduce a wider range of area measures and the concept of conservation of area, provide a large number of decimetre squares (cut from plain paper) and suggest that the children paint them, cut a square metre of paper, then use the square decimetres to cover the square metre.

- How many did they use?
- Did they make a pattern?
- How many centimetre squares are there in a square metre?

On another occasion, introduce millimetre squares. Ask the pupils how many millimetre squares there are in a square metre. Then ask them to cut a square metre of millimetre square paper. They can label it: 'How many squares are there here?'

Understanding the relationship between area units: finding perimeters

For Key Stages 2 and 3, from time to time revise the set of relationships between area units from a square millimetre to a square metre. Question the pupils about the ratio of successive area units (1/100). Ask them if they can find a space in the school which is 100 square metres in area. This is called an **are**. (100 ares make one hectare. What is the length of its side?)

The two extensive problems which follow can be solved by pupils at Level 4 by a practical method. Pupils at Levels 5 and 6 can also solve the problem by making algebraic graphs. Co-ordinators should introduce them in INSET sessions first.

Activity 3.25 Ask the teachers to investigate the areas of two-dimensional shapes with equal perimeters. Can they change the area?

Ask them to cut some strips of graph paper 20 cm long, 5 mm wide, and to join the ends of each with Sellotape, without overlap. As they put these on the table, what shape do they take up? Does this suggest the shape of maximum area?

Ask the teachers to experiment with the loops, making them into different shapes. They will have to fold the 20-cm length into 4 or 5 equal lengths first. Then, using a sheet of centimetre squared paper, ask them to count the area enclosed by the loops.

Areas of regular shapes will be easier to count, e.g. equilateral triangle, square, regular pentagon, etc, but they will have to find an approximate area for many of the loops, e.g.the circle.

Then ask: 'What is the smallest area you can enclose?' (A 9-year-old said, 'Look, I've squashed out all the area!')

Activity 3.26 Ask the teachers to investigate the areas of rectangles of perimeter 20 cm. They can begin by making an ordered table: width, length, area (cm). Ask them to describe the number patterns. Then ask them to draw first a w/l graph, then w/area graphs. Do they recognize the w/l graph? What is the algebraic equation of this graph? ($w + l = 10$.) Did the teachers remember the line of 7s in the addition graph? ($x + y = 7$.)

Then ask them to cut four complete sets of the rectangles contained in their table. They can arrange three of these sets in order in different ways, as shown in Figure 3.3 (a–c). Ask them to describe each set. (In the third set the rectangles should overlap but

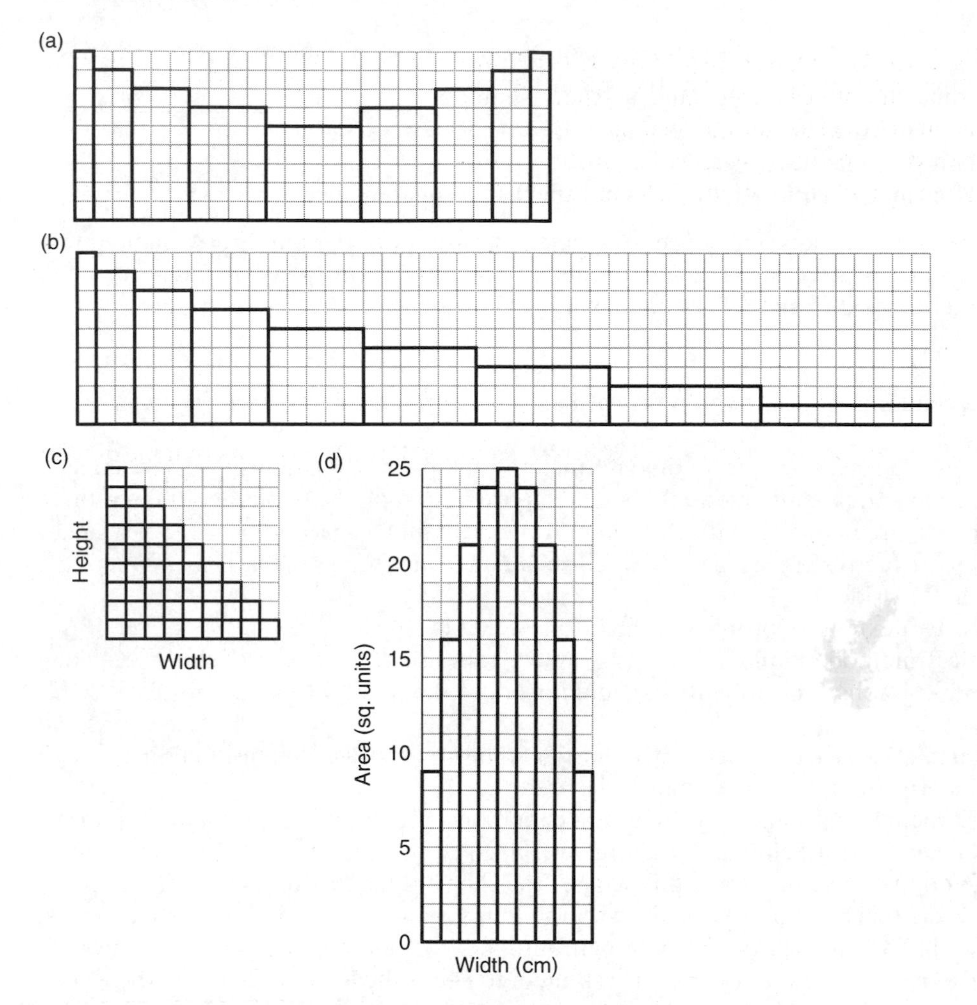

Figure 3.3 *Investigating areas of rectangles of perimeter 20 cm: (a) arranged with mirror symmetry; (b) arranged in order of decreasing height and increasing width; (c) overlapping rectangles; and (d) areas represented by a block graph (sq cm)*

the order of 3.3(b) should be kept.) The fourth set of rectangles should be used to make an area/width graph. Cut each rectangle in turn into centimetre strips and join these strips to make one long strip representing the area of that rectangle. Then construct a block area graph with them (Fig. 3.3(d)).

What is the shape of the rectangle with the largest area? What is the smallest area? Did the teachers suggest zero?

We can use symbols w, l, and A, where $w + l = 10$, and $A = wl$, giving $l = 10 - w$ and $A = w(10 - w)$ (or $y = x(10 - x)$ if preferred).

w	l	A
1	9	9
2	8	16
3	7	21
4	6	24
5	5	25

The second problem is the reverse of the first. This time the area is fixed, and teachers should investigate the perimeters of rectangles of area 36 square centimetres. In your INSET group we suggest you use rectangles of area 16 cm^2.

w	l	A
1	16	16
4	4	16
3.2	5	16

Activity 3.27 Begin by making a table of width/length/perimeter (Table 3.2).

Table 3.2 *Investigating rectangles with area 16 cm^2*

Your entries could be:								
	width (cm)	1	2	4	5	8	10	16
	length (cm)	16	8	4	3.2	2	1.6	1
	perimeter (cm)	34	20	16	16.4	20	23.2	34

Ask the teachers to cut three sets of the rectangles in this table (from cm mm squared paper) and to arrange each set of rectangles in order of width as in Fig. 3.4 (a-c). The third set should be arranged in order but overlapping as before, and this graph (Fig. 3.4(c)) should be familiar to the teachers – the multiplication square may remind them of this relationship.

Now, look at the numbers in the first and second columns of the table. When the width is doubled, what happens to the length? When the width is quadrupled,

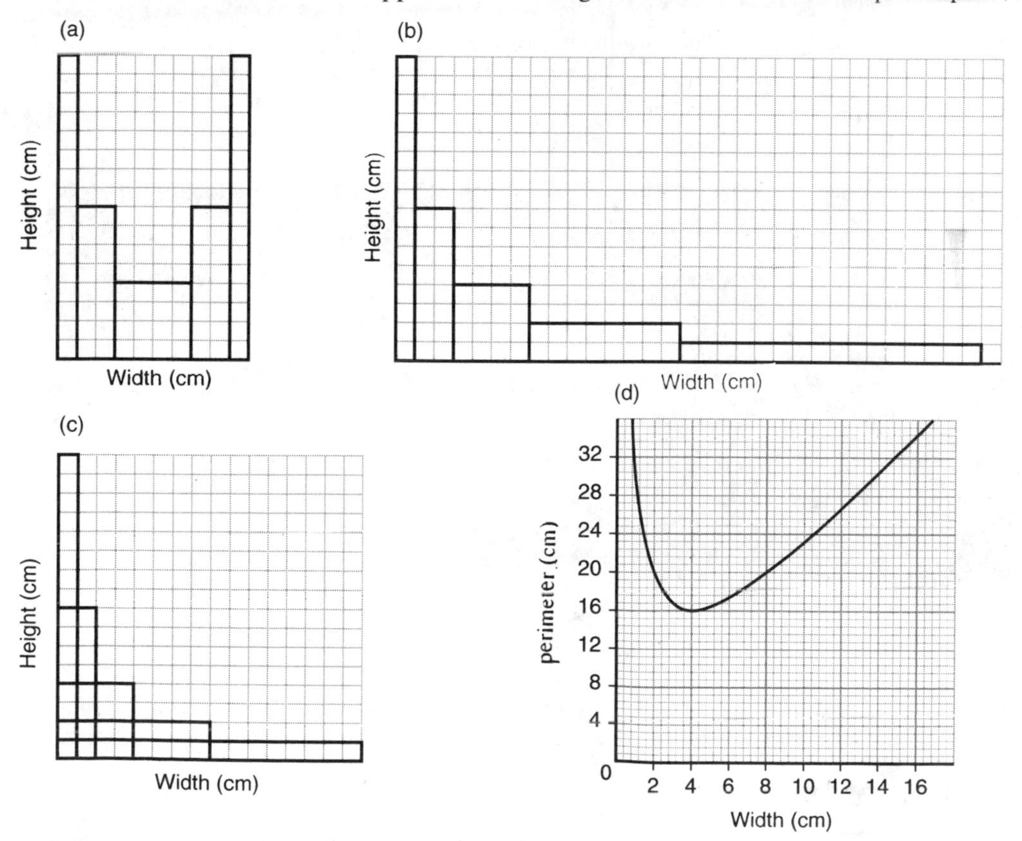

Figure 3.4 *Rectangles of area 16 m^2: (a) symmetrical arrangement of cut rectangles; (b) arranged in order of increasing width, decreasing length; (c) rectangles overlapping, $w + l = 10$; and (d) width/perimeter graph*

what happens to the length? And vice versa? Does this suggest the equation of the width/length function? What does the product tell you? If you introduce this problem in class, remember to include all the steps, although a fixed area of 36 (which has more factors) would be easier for pupils to work with.

At Key Stage 3, pupils should also draw the width/perimeter graph. Ask them if they can find out whether the graph (Fig. 3.4(d)) is symmetrical. Also ask them what happens to the length as the width gets shorter, e.g. $w = 0.1$, $l = 160$; $w = 0.01$, $l = 1600$; and so on.

Area of a circle

At Key Stage 3, the areas and perimeters of various plane shapes are considered in Chapter 5 but for completeness we include the area of a circle here.

Give each group of pupils a metre of picture cord. Ask them to halve it and to coil the two pieces tightly so that they make two identical circles. Then ask them to stick one of the two coils onto a piece of card (as in Fig. 3.5(a)). The second should be glued underneath the other but before the glue sets they should cut through the strands along a radius (as shown in Fig 3.5(b)) to create a triangle of circumferences.

Ask:

- What is the length of the base of the triangle? ($\pi \times 2r$)
- What is the height of the triangle? (r)
- What is its area? ($1/2 \times 2\pi r \times r$, or πr^2)
- So, what is the area of the circle? (πr^2)

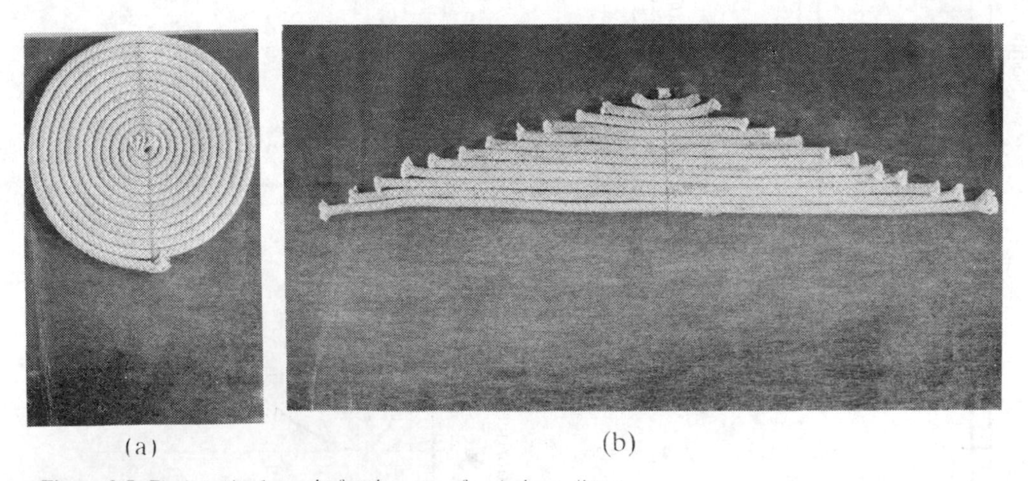

(a) (b)

Figure 3.5 *Finding the formula for the area of a circle, radius r*

Give your pupils plenty of practice in applying this formula for the area of a circle.

TIME (KEY STAGE 1)

Estimating the passage of time is difficult for adults as well as children. Our estimate depends upon our mood as well as on the activity we are involved in. Sometimes an hour can seem like a day and yet years pass very quickly as you grow older!

The vocabulary children have to learn for 'general' time is extensive:

today tomorrow yesterday
last week next year soon
sometime never

Another concept with its own vocabulary associated with time is speed:

slow fast slowly quickly

Ordered sequences also have to be memorized: the names of the days of the week and of the months of the year; the number of days in each month. Later on, geological time and historical time will be introduced.

One of the most important concepts involving time is duration of time. Many daily activities depend on the ability to calculate duration of time: journey times; cooking times; watching TV; concerts and theatre performances; programming a video recorder, using a microwave, or a washer/dryer. Many local bus and train timetables are available to provide pupils with regular practice in calculating duration of time.

Reading the time

Unfortunately, there are still a few pupils who are unable to read the time when they reach the secondary school. It is not an easy skill to acquire because the time of day can be expressed in several different ways.

Analogue clocks versus digital clocks

Analogue (circular) clocks record the time in 12-hour blocks. The time can be read as 'past the hour' or 'to the hour'. For example, 25 past 3 or 3.25; 25 to 4, or 3.35. The quarter hours and half hours are also used to denote time on these clocks.

Train and plane timetables work on 24 hours. Many stations now have 24-hour digital clocks. These record only time past the hour.

Reading the time should eventually include reading with understanding a digital watch display, a traditional circular watch (12 hours), and a 24-hour station clock.

First, the children learn to read the hours, half-hours, quarter-hours and three-quarter hours on the circular clock. Then we need to make sure that they can count in 5s to 60, as well as tell us how many fives there are in these multiples of five. The following activity allows them to make their own 'clock'.

Activity 3.28 Ask pupils to make a number line on centimetre squared paper, 24 cm long and 1 cm wide, labelled in fives between the numbers 1 to 12 (at 2 cm intervals). Join the ends without overlap, with the numbers facing inside, using Sellotape.

Then cut out a circle of diameter 8 cm from thin card to fit inside the number line. (Before the children fit the circle within the number line, ask them to find and mark the centre. How did they do this – by folding?) The pupils should mark the corresponding hours (1 to 12) clearly, in another colour. The pupils can also make movable hands, of differing lengths, to fasten to the centre of their individual clock faces.

(In your INSET group, suggest that the teachers make a number line to encircle the perimeter of their classroom clock – approximately 20 cm in diameter, 63 cm in circumference, with an overlap of 3 cm, 12 units of 3 cm, labelled in red from 0 to 12 for the hours, and 0, 5, 10, etc. for the minutes to 60.)

Each day, for at least a couple of months, give groups of pupils practice in reading the time to the nearest five minutes. Sometimes ask the children to show different times on their own clock faces.

Activity 3.29 Investigate the number of hours during which the children you teach are in bed. Ask their parents to supply the information, to the nearest hour, for the previous evening, and the current morning. Discuss the kind of block graph the pupils want to make. Ask them what information they want to show. What range of hours should they show?

Ask the children:

- Who goes to bed earliest?
- Who goes to bed the latest?
- Who gets up first?
- Who is in bed the longest?

The graph shown in Figure 3.6 was made by 7-year-olds.

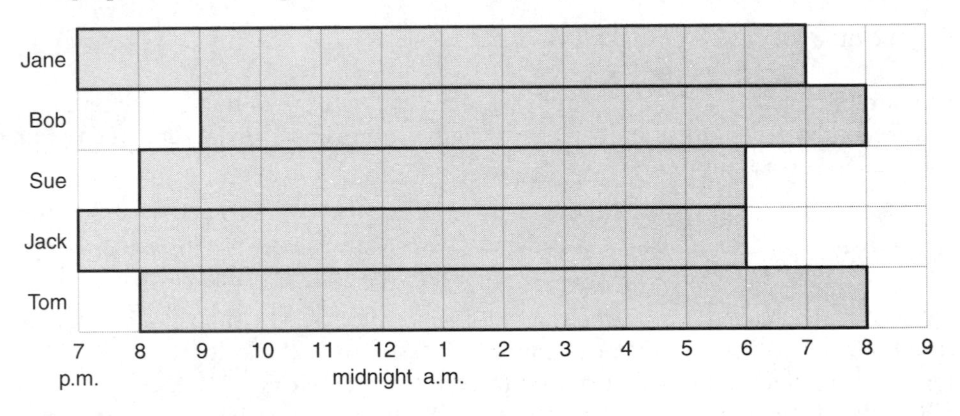

Figure 3.6 *How long are we in bed?*

Dates

Activity 3.30 Keeping a weather record for two months is a useful activity providing experience of writing different dates. The pupils can transfer the information daily to a weather calendar. The first record can be simple, with entries 'sun, cloud or rain'. At the end of the month the pupils can find out the most and least frequent type of weather.

On another occasion the pupils can make and use a rain gauge. They need a cylindrical tin about 20 cm high and 10 cm in diameter. This is fitted with a 10-cm-diameter funnel. The teacher should collect seven identical clear plastic bottles, about 2 to 3 cm in diameter. At the end of each day the rain collected in the tin will be poured into a bottle labelled with the day of the week. The rain in the bottles forms an instant graph. Ask the pupils to describe the graph.

Later on, the pupils can learn to read the wind direction from a weather vane and relate this to the rainfall. They may also be interested to compare the class records with those in the local newspaper.

Making 'timers': sand timers, pendulums, shadow clocks (Key Stage 2)

Making sand timers

Groups of pupils will need two identical small plastic bottles (about 100-ml size), joined by a single cork with a hole bored through it. Make sure that for a sand timer the sand is fine and dry. Let the pupils use a time clock or a watch with a second hand to find the time the sand takes to flow from one bottle to the other. Suggest that they try to make this into a minute timer. They can make a list of tasks they can complete in one minute.

Making a pendulum

The pendulum can be made of a flattened lump of Plasticine fastened to one end of one metre of fine string. It is important to find a good point of suspension. The other end can be fastened by a strong drawing-pin to a point in the lintel over a door.

Case 3.9 The children were asked to set the pendulum swinging by starting it level with the point of suspension. Then they were asked to find out whether the pendulum beats regularly, and to describe how they did this. One group timed ten complete swings from one extreme position and back again. They repeated the timing for three successive sets of ten swings. Another group counted the number of swings completed in successive minutes for four minutes in all. Both groups decided that the swings seemed to be regular. But they noticed that the pendulum finally came to rest!

Then the teacher asked them how they could change the beat. The suggestions came quickly: add a bit of Plasticine on, take a bit off; shorten the string, lengthen it; start the bob somewhere else; give it a push! This investigation took some time, and only two of the suggestions changed the beat.

Shadow clocks

These investigations are fun and informative – and all that is needed is a sunny day!

Case 3.10 The teacher of a reception class had taken the children outside to fix a shadow stick in the playground. At 9 a.m. Joanna used chalk to draw the shadow, and the teacher wrote 9 a.m. at the end of the shadow. Mark, the youngest in the class, was asked to draw the shadow at 9.30 a.m. He waited anxiously for the clock hand to reach 9.30.

Shortly afterwards an angry boy shouted at the teacher: 'You moved it! I'm going to watch you now so you can't move it again.' The teacher was hearing Joanne read and did not realize at once what was the matter. When she did, she took the children out to sit and watch the moving shadow.

Later on, when all the shadows had been drawn at half-hour intervals, the children were excited to see that the shadow was shortest at midday.

Case 3.11 Children at Key Stage 2 investigated the shadows of three-dimensional objects. A 1-litre plastic bottle, which they weighted down with sand, was placed in the middle of a large sheet of paper, in the middle of a quadrangle, and they drew round the base in case the bottle was shifted. They then drew round the shadow.

When Patsy and Bob came out half an hour later to draw the second shadow, they found that their apparatus was already in shadow. For the rest of that day the pupils looked out for a position which would be clear of shadow until the end of the afternoon. They decided to position the apparatus in the field and to begin again next day.

The finished shadow picture was labelled with the times of observation. The shortest shadow was easily seen to occur at 12 noon.

Investigating speed (Key Stage 2)

Case 3.12 The children rigged up a plank to act as an incline. They fixed one end of the plank on four bricks to begin with and started a truck from rest at the top. Eventually they had to fit rails to the plank because the truck kept falling off the plank. They used a stop clock for timing.

The same boy started the clock and stopped it when the front of the truck reached the floor. They recorded the time taken for the truck to travel down the ramp five times. Then they repeated the experiment with the ramp on 3 bricks, then 2, then 1. Later on, they decided to load the truck to see if they could slow it down.

The teacher then began to question the children about their own walking speed. She asked them how long they would take to walk a kilometre. No one knew. She asked if anyone knew how far they lived from school. George thought that he might live a kilometre from school. 'It takes me twenty minutes to walk – but I don't hurry.'

The teacher then questioned them about a kilometre. How many metres was it? 'A thousand metres', George said, 'because kilo means thousand.'

The children asked if they could mark a kilometre running track on the field and organize speed trials for everyone to take part. Jane pointed out that if everyone took 20 minutes, the trials would take a long time. So they decided to begin by walking 100 metres at a normal walking pace. They appointed a starter with a flag, and two at the finishing post, to record the number of seconds each pupil took to walk 100 metres.

The times taken to walk 100 metres ranged from 59 seconds to 120 seconds. The results were posted in order on a notice board. All the children worked out how long it would take them to walk a kilometre at the same rate. 'But could you keep up that pace for a kilometre?' George asked. Most of them thought that they could because they had been walking, not running. Some of them decided to find their running speed over 100 m. Others found their cycling speed over the same course.

The teacher then asked if the children knew another method of expressing speed. 'What about the speed of aircraft?' she asked. 'Concorde is supersonic', said George. The teacher explained that this time, instead of finding how long they took to cover 1 kilometre, they needed to find how far they travelled in a set time. They agreed on a minute. 'Then we could work out how far we could travel in an hour, at the same rate.' 'If we could keep it up', said George.

A graphical record of time/distance problems will be found in Chapter 6.

MONEY

First notice whether or not pupils are recording sums of money correctly. Dictate various amounts of money and ask the pupils to record these in a list. Notice which pupils record these sums of money correctly. Show the other pupils the right way to record sums of money and give short daily practices for those who need it. (NB £1.52p is incorrect.) Also take the opportunity to provide experience of recording, in decimals, amounts such as £0.5 which is better recorded as £0.50, to distinguish it from £0.05 or 5p.

Then make sure that pupils can say aloud amounts of money in decimal form. Sometimes ask them to state these in pence; e.g. £0.60 as 60 pence, £0.10 as 10p, £0.01 as 1 penny, £1.75 as 175 pence, etc.

According to the level reached by the pupils dictate examples such as:

Find the total amount Mrs Brook spent in one day: £1.52, £13.75, £0.30, £9.70.

The topic of money is a valuable one for applying the four operations. Calculations involving money have already been included from time to time e.g. the shopkeeper's method of giving change, when discussing subtraction. Here are yet more examples.

Activity 3.31

1. Mrs Black shops at two supermarkets. One day she spends £12.45 at one and £15.23 at the other.
   - How much did she spend altogether?
   - What was the difference between the two bills?
   - What change did she have from £30?

   Encourage pupils to use the shopkeeper's method of subtraction. (Mrs Black spent £27.68 altogether, and her change was £2.32.).
2. Mr Black buys 6 bottles of cola at £2.65 each. What was the cost and how much change did Mr Black get from £20?

   2 bottles cost £5.30
   6 bottles cost £15.90
   Change from £20 is £4.10.
   Check: £15.90 + £4.10 = £20.
3. How many calculators at £3.85 each can a school buy for £60? Division can often be more easily done by multiplication.
   10 calculators cost £3.85 × 10, or £38.50
   5 calculators cost ½ of £38.5 or £19.25
   15 calculators cost £57.75
   Change £2.25

So 15 calculators can be bought for £60.

It is important to keep prices up-to-date. Ask the pupils to bring in some recent catalogues. Sometimes let them make up examples for each other. Encourage the pupils to discuss and appraise methods.

Another attractive source of problems involving money can be centred on holiday brochures. These should also involve journeys, and the total planning for the family on holiday – at home or abroad.

Chapter 4

Algebra

INTRODUCTION

Algebra is concerned with pattern-making and relationships. It helps us to express concisely particular relationships we discover in number, to generalize these relationships and to use symbols to represent them. It also utilizes another powerful tool – graphical representation – from which we can make further discoveries. Once pupils appreciate the tools offered by algebra, their powers of problem-solving are greatly increased.

Here is a frivolous example. This problem involves two equations, called simultaneous equations.

> The cost of 3 doughnuts and 2 buns is 84 pence. The difference between the cost of a doughnut and a bun is 8 pence. Find the cost of each.

If d and b pence are the cost of a doughnut and of a bun we can write:

$$3d + 2b = 84 \qquad (1)$$
$$d - b = 8 \qquad (2)$$

if we assume that doughnuts are dearer than buns. Now we need to create an equation involving b or d only. Can you see how to do this? If we double equation (2) we find:

$$2d - 2b = 16 \qquad (3)$$

If we add equations (1) and (3) the terms '$+2b$' and '$-2b$' cancel out and we have

$$5d = 100$$

from which we find $d = 20$. From equation (2), we then find $b = 12$. (Check from (1): $60 + 24 = 84$.) The cost of a doughnut is 20 pence and that of a bun is 12 pence. If we had not availed ourselves of algebra we might have taken much longer to solve this problem.

Some primary teachers did not enjoy algebra at school – perhaps because, at the time, they did not see any reason for learning the subject. If our pupils are to enjoy learning algebra and to appreciate its value in solving problems – as the ancient Egyptians did more than 5000 years ago – they must see their own teachers enjoying this subject. Let's make it fun!

INTRODUCING PATTERNS

The Programmes of Study at Level 1 suggest an attractive beginning to algebra: pupils should be copying, continuing and devising repeating patterns, using objects, apparatus and single-digit numbers.

Activity 4.1 Teachers, too, should begin by making and describing repeating patterns:

- a bead necklace (red, blue, blue, red, blue, blue, . . .);
- a potato print;
- a train made of interlocking cubes (yellow, green, yellow, green, yellow . . .);
- a chain of numbers: 3, 5, 3, 5, . . . ;
- a line of shapes.

In the classroom, encourage the children to make patterns of any repetitive material they find: for example, coloured rods of two or three different lengths. They should describe their patterns and develop them. This activity is valuable in helping children to recognize, for instance, patterns of 10 without counting one by one. The patterns of 10 below were made and described by 6-year-olds. Try to encourage children to describe each pattern in many different ways.

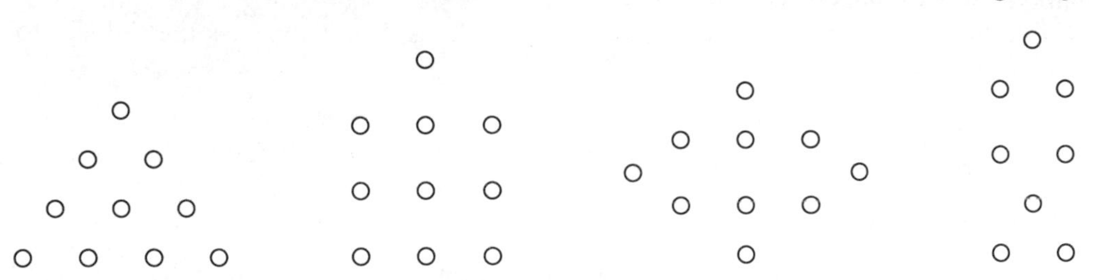

At Key Stage 1, pupils should explore number patterns:

- distinguishing odd and even numbers;
- understanding the use of a symbol to stand for an unknown number.

The children can suggest other examples of number patterns; provide counters for them to illustrate the patterns:

$6 + 0 = 6$	$6 = 5 + 1$	$7 - 4 = 3$
$5 + 1 = 6$	$6 = 4 + 2$	$7 - 3 = 4$
$4 + 2 = 6$ etc.	$6 = 3 + 3$ etc.	$7 - 2 = 5$ etc.

Odd and even numbers

Some children will already recognize odd and even numbers. Give them daily practice for a few weeks, with questions such as:

- Is there an odd or an even number of children here today?
- How do you know?

Do they join hands with a partner? Ask:

- Is anyone odd man out? (without a partner)
- Is there an odd number of girls?
- Is there an odd number of boys?

Encourage the children to record any findings concerning the odd and even numbers they discover.

Symbols standing for unknown numbers

Activity 4.2 In your INSET group, ask the teachers for examples of missing numbers which children could be asked to find. Would their children recognize any of these?

1, 3, 5, *, 9
2 + * = 10
* + 2 = 7
6 − * = 2

Do the children find missing number problems easy? If so, ask them to make up some examples for the class book of problems. It is useful to produce a 'fancy' box to hold the missing number, perhaps a cube. Put the box in place in each sequence in turn. Encourage the children to guess the missing number, and to check, saying the sequence aloud, perhaps backwards, e.g. 9, 7, 5, 3, 1.

Give them some oral 'Think of a number' problems. Every time ask the children to explain their method.

- I added 7 to a number and the answer is 19. What's my number?
- I subtracted 3 from my number; the answer's 17. What's my number?
- When I halved my number the answer's 9. What's my number?
- When I trebled my number the answer is 39. What's my number?

Remind them of what you did, ask again what they did, until they realize that they have to 'do the reverse'. For example, if you add 7, they have to subtract 7 – you subtract 3, they add 3 – you halve a number, they double the number quoted.

Sometimes ask the children to check their answer and show you how they found it. Give daily practice in 'Think of a number' problems until the children achieve a high standard of accuracy. (Keep to whole number answers.)

Developing strategies to perform mental calculations, using number patterns

At Key Stage 2 the emphasis continues to be on the use of pattern, particularly when doing mental calculations, e.g.

- continue 5, 10, 15, 20 . . .
- also, 10, 20, 30 . . .

How many 10s are there in 70, in 90, etc.? Also prepare for halving 90, 70, . . . (This will help children to find how many 5s there are in 45, etc.) Can they continue the following pattern?

$7 + 10 = 17$
$17 + 10 = 27$
$27 + 10 = 37$

Give mental problems such as: $26 + 53$. Do they rewrite the sum as:

$20 + 6 + 50 + 3 = 70 + 9$?

Give plenty of examples. Use inverse operations in a simple context: e.g. doubling and halving, adding and subtracting. Ask the group to recall all the inverse operations they have met so far:

- add and subtract
- subtract and add
- multiply and divide
- divide and multiply.

Ask if anyone remembers other inverse operations. Some may remember squaring a number, say $5 \times 5 = 25$, and then finding its square root.

Introduce teachers to the idea of a function machine.

Input	Output
0	3
1	4

Ask questions such as: What is the operation? Sometimes reverse input and output and ask the same question. Also, include some examples from LOGO (see Chapter 9).

Back in the classroom, question the children about inverse operations. Which operation undoes addition? division? subtraction? multiplication? Remind them about using inverses to check calculations. Then give examples such as:

$5 + 4 = ?$ (9)
How can you get back to 5? (Subtract 4 from 9 or $9 - 4$.)
$7 \times 4 = ?$ (28)
How can you get back to 7? (Divide 28 by 4.)

Provide frequent short practice.

Make function machine operations fun! (Perhaps hold a competition for the most interesting function machine – and later on, for the most interesting – and amusing? – operation.) Provide examples using all four operations. Suggest that your pupils make up examples to give to the class.

BEGINNING TO GENERALIZE

At Key Stage 2, for the first time, pupils are expected to generalize, mainly in words, patterns which arise in various situations, e.g. symmetry of results:

$3 + 2 = 2 + 3$
$4 \times 6 = 6 \times 4$

Applying strategies such as doubling and halving to explore properties of numbers, including equivalence of fractions

In former times, multiplication was often achieved by doubling one number and halving the other.

$37 \times 8 = 74 \times 4 = 148 \times 2 = 296 \times 1 = 296$

Activity 4.3 Use the same method of doubling and halving to calculate:

(a) 157×8
(b) 225×16

Activity 4.4 Provide identical sheets of paper and ask each pupil to fold their clean sheet into 8 identical parts. Ask them to use this to find different equivalent fractions for $1/2$, and they will probably suggest $2/4$ and $4/8$.

If asked to write another five fractional equivalents for $1/2$, they may suggest $3/6$, $5/10$, $6/12$, $7/14$ and $8/16$. Ask how they know that all these fractions are equivalent to $1/2$. The usual reply is that they all 'cancel' to $1/2$, but what do we mean by 'cancel'? Someone may say, 'I multiplied the numerator and denominator of $1/2$ by the same number.' Then ask, 'So what operation did you have to use to obtain $1/2$ once more?' (Division)

In an INSET session, ask each teacher to choose a different simple fraction (such as $1/3$, $1/5$, $1/7$ . . .) and to make a string of equivalent fractions for it. Challenge the teachers to think of a problem to illustrate the concept of equivalent fractions which they can put to good use in the classroom. Here is one example, but listen to teachers' suggestions first.

Activity 4.5 Lengths of $1\frac{1}{5}$ m and of $1\frac{1}{10}$ m were cut from $3\frac{1}{2}$ m. What length is left? All these fractional lengths can be expressed as tenths:

$\frac{1}{5} = \frac{2}{10} \quad \frac{1}{2} = \frac{5}{10} \quad \text{and} \quad 1 = \frac{10}{10}$

So

$$1\tfrac{1}{5} + 1\tfrac{1}{10} = 1\tfrac{2}{10} + 1\tfrac{1}{10}$$
$$= 2\tfrac{3}{10}$$

The length left is calculated by subtracting $2\frac{3}{10}$ from $3\frac{1}{2}$, i.e.

$3\frac{5}{10} - 2\frac{3}{10} = 1\frac{2}{10}$

Since $2/10 = 1/5$ we can simplify the answer to $1\frac{1}{5}$m.

PATTERNS AT KEY STAGE 2

Activity 4.6 Ask the pupils to make addition (Fig. 4.1) and multiplication (Fig. 4.2) tables. In both tables look for number patterns and describe these in as many different ways as possible. Start with the major diagonal on the addition table, and describe the pattern on it. Encourage complete answers, e.g.

- multiples of 2 from 2 to 12,
- the 2 times table (from 2 to 12),
- even numbers from 2 to 12,
- doubles of numbers 1 to 6.

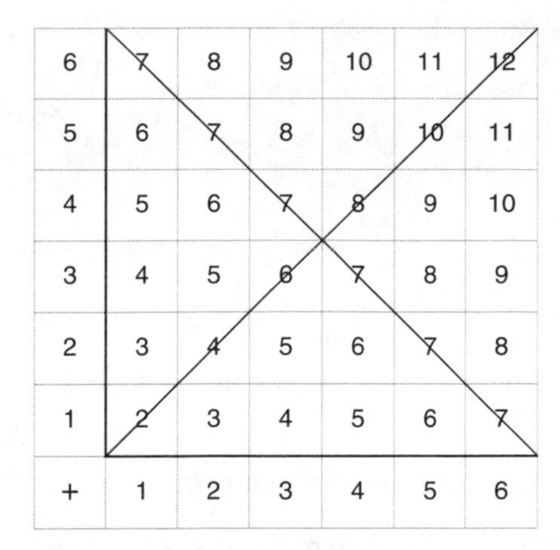

6	7	8	9	10	11	12
5	6	7	8	9	10	11
4	5	6	7	8	9	10
3	4	5	6	7	8	9
2	3	4	5	6	7	8
1	2	3	4	5	6	7
+	1	2	3	4	5	6

Figure 4.1 *Addition table*

Ask why these numbers are multiples of 2, or doubles of 1 to 6.

Discuss the numbers on the second diagonal of the addition table. (Later on, the teachers will find an algebraic way of describing this line of 7s: $x + y = 7$.) What does this line of 7s mean? (There are six different ways of making 7: $6 + 1 = 5 + 2 = \ldots$ etc.)

6	6	12	18	24	30	36
5	5	10	15	20	25	30
4	4	8	12	16	20	24
3	3	6	9	12	15	18
2	2	4	6	8	10	12
1	1	2	3	4	5	6
×	1	2	3	4	5	6

Figure 4.2 *Multiplication table*

Treat the multiplication table in the same way. Pay special attention to multiples of 2, 5, and 10. Introduce the concept of multiples and factors functioning as inverses of one another: i.e., 60 is a multiple of 10; 10 is a factor of 60. Why are there squares on the diagonal?

Two-stage problems

The work on function machines can be extended to two-stage problems, e.g. a machine which multiplies by 10 and adds 5.

Activity 4.7 Complete the table for a machine which divides by 5 and subtracts 6.

Input	Output
55	5
60	..
65	..
70	..
75	..
80	10

This provides a useful introduction to oral two-stage problems, of the 'Think of a number' type, e.g.

- When I doubled a number and added 2, the answer was 20. What is the number?

Encourage the pupils to describe how they solved the problem, particularly the order of operations. Provide examples using every operation (addition, subtraction, multiplication and division), two at a time.

Understanding and using simple formulae – algebra as a shorthand

Algebra is often described as the shorthand of mathematics – a way of expressing relationships – as the following examples illustrate.

Example To find the area of a rectangle we multiply its length by its width, and this can be written more briefly as

Area of a rectangle = length × width

More briefly still, if we suppose its length is l cm and width w cm, where l and w can be represented by any numbers we choose, whole or fractional or mixed, then its area A in cm^2 is given by

$A = l \times w$ or $A = l.w$

where . means multiply, or even more briefly, as

$A = lw$

where we understand that lw means $l \times w$. This formula gives concise and clear instructions for finding the area of a rectangle.

Example Similarly, we can find a formula for the perimeter, p cm, for this rectangle:

$p = l + w + l + w$

or

$p = 2l + 2w$

where $2w$ is taken to mean $2 \times w$.

Activity 4.8 Find a formula for the area, A cm^2, of a square with sides s cm. Then find a formula for its perimeter p cm.

The answer $A = ss$ can be written more concisely as $A = s^2$. (Where $ss = s \times s = s^2$, called 's squared'.) If V is the volume of a cube of edge c cm the short way of writing $V = ccc$ is $V = c^3$ and is read: the volume is 'c cubed'.

The children will need regular practice in finding and using simple formulae, and in writing these in as simple a form as possible – collecting like terms and writing $w + w$ as $2w$, ss as s^2, ccc as c^3, and so on. You should however, emphasize that we can only collect like terms; e.g.

$$\begin{aligned} 2a + 3b + 5c + 4a - 2b &= (2a + 4a) + (3b - 2b) + 5c \\ &= 6a + b + 5c \end{aligned}$$

where the brackets tie the like terms together. The next activity provides some examples.

Activity 4.9 Find the number n of metres in C cm. (Ans: $n = C \div 100$.) Then find shorthand forms for the number n of:

(a) metres in k kilometres
(b) seconds in m minutes
(c) hours in d days
(d) pence in P pounds
(e) cm in M metres
(f) grams in K kilograms
(g) hours in M minutes
(h) kilograms in G grams
(j) days in H hours
(k) kilometres in m millimetres

Activity 4.10 Find the distance Jenny walks in 2 hours at 5 km per hour. Then find a formula for the distance k kilometres travelled in h hours at a speed of s km per hour.

Activity 4.11 The six members of the Wood family pay £N altogether for theatre tickets. What is the cost of one ticket?

Activity 4.12 The perimeter of a square room is P metres. What is its length?

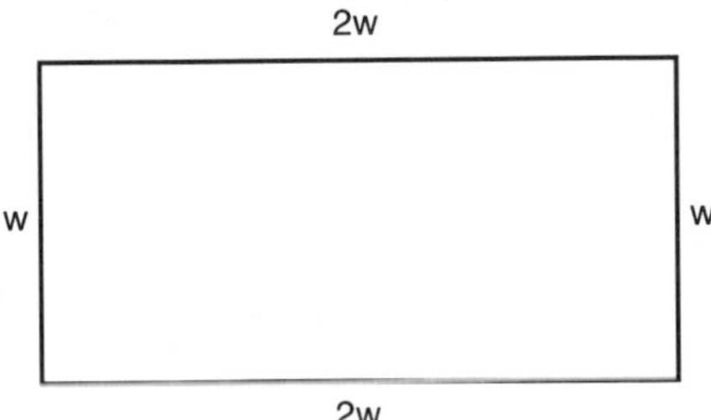

Figure 4.3 *Sheep pen, w by 2w*

Activity 4.13
Figure 4.3 shows a rectangular sheep pen. Find a formula for the length f metres of fence, in terms of w, the length of the shorter sides. Use the figure to find the area of the sheep pen A (m^2). Find the perimeter and area when (a) $w = 15$ m and (b) $w = 20$ m. Use calculators.

Activity 4.14 In Chapter 5 you will find that for a circle of radius r cm, circumference C cm and area A cm^2, $C = 2\pi r$ and $A = \pi r^2$. Taking $\pi = 3.14$, calculate C and A when (a) $r = 4$ cm, (b) $r = 7.2$ cm (give your answer to 3 sig.fig.).

Activity 4.15 The formula for converting I inches to C cm is: $C = 2.54 \times I$. Find the number of centimetres in 10 and 100 inches (to the nearest half-centimetre).

The formula for converting temperatures from Fahrenheit to Celsius scales is $C = \frac{5}{9}(F - 32)$. Find the Celsius equivalent of (a) 41°F, (b) 50°F, (c) 68°F.

Learning the conventions of the coordinate representation of points: working in the first quadrant

Activity 4.16 Arrange the desks or tables in your classroom in rows and columns within a rectangular block as far as possible. Then ask each pupil to write a description to enable a friend in another class to find the writer's desk. Ask some of the pupils to read their description aloud to the class. If there is any doubt about the correctness of their description, ask another pupil to try out the instructions.

At the end of the exercise, ask the pupils what all the descriptions had in common. Have they all counted the desks from two walls at right angles? Ask them to draw two walls (OX and OY) at right angles (as in Fig. 4.4), and label OX and OY as shown, and number the points (for desks) from 0 as in the figure. Give the pupils a set of points to plot to make a picture, e.g. an octagon.

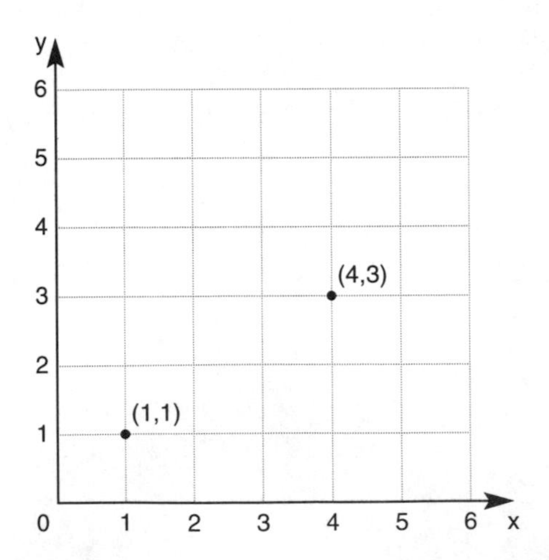

Figure 4.4

When they have plotted all the points and joined them as instructed, ask them to name the picture. Explain that to avoid confusion we always state the number on the horizontal axis (OX) first, and secondly the number on the axis (OY), called the vertical axis.

Activity 4.17 Plot the points on a new set of axes (marked 0 to 12 on both axes) at right angles, joining each point to the next.

(0,9), (2,10), (4,9), (9,9), (11,7), (11,0), (10,0), (9,4), (8,4), (8,0), (7,0), (6,4), (5,4), (6,1), (5,0), (4,2), (3,0), (2,0), (3,5), (2,5), (1,3), (1,1), (0,2), (0,9)

Next make an eye at (1,7); an ear (2,9), (3,7), (4,9); a tail (11,7) to (12,4). The finished picture should look like Figure 4.5!

Then ask the pupils to draw another set of axes and to make their own picture. They can make a class book of problems – a set of points to plot, with instructions for joining points to complete the picture.

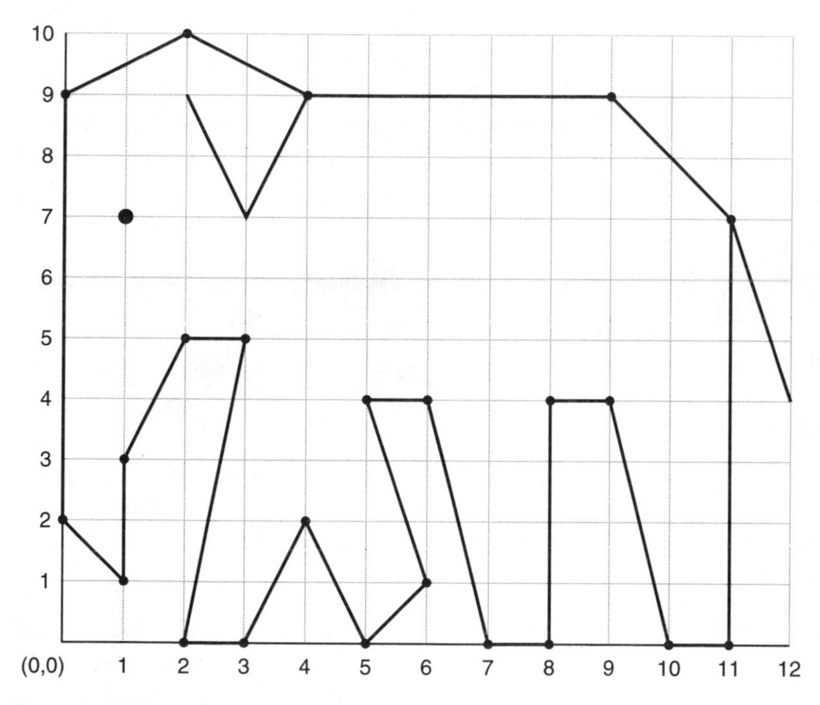

Figure 4.5 *A dotty elephant*

Vocabulary O is called the *origin* (beginning). The two numbers which describe the position of any point, for example, (3,4), are called *coordinates*; the first number refers to the 'horizontal' axis, the second to the 'vertical' axis.

Understanding simple equations

At this level equations can be expressed in words, and these can arise from 'Think of a number' problems, e.g.

> I doubled a number, subtracted 6 and the answer is 12.
> What's my number?

KEY STAGE 3: SEQUENCES

Generating sequences, by following instructions and by recognizing and continuing spatial patterns.

To generate a sequence we need to know the starting point and the relationship that will allow us to calculate all subsequent terms in the sequence, e.g.

Starting with 1, each successive term is double the previous term

will generate the sequence 1, 2, 4, 8, 16, . . .

In an INSET session ask the teachers to work in groups to generate a variety of sequences. Give them a simple example such as the one above, and ask for the same sequence, but with an earlier starting point. Does anyone suggest starting with $1/2$? If not, ask for the term before 1. What about 0? Could 0 belong to this sequence?

One group may suggest a sequence which grows smaller, rather than larger, such as: 1, $1/2$, $1/4$, and so on. What instructions do they give? What would be the term before 1?

Make sure that equations made by addition and by subtraction are also included, e.g.

Term		Pattern?
1st	1	1
2nd	4	$(1 + 3)$
3rd	7	$(1 + 2 \times 3)$
4th	10	$(1 + 3 \times 3)$
5th	13 . . .	. . .
. . . 10th . . .		
nth (any term)		

Ask them to discuss the pattern. What about the number of 3s for each successive term? They should notice that there is one fewer than the number of the term, i.e. the **5**th term is ($1 + \mathbf{4}.3 = 13$). Ask how many 3s the nth term will have ($n - 1$). So the nth term is $1 + (n - 1)3$.

Activity 4.18 Provide small pebbles, shells or buttons and ask the teachers to use these to generate sequences. Then ask them to write instructions for generating the sequences. Make sure that the first term is always fully discussed. For example, does the first term in a sequence of triangular numbers have to be 3? Were squares, pentagonal and rectangular numbers included in the sequences? Did you always consider whether 0 could be a first term? Can they write a formula for the nth term?

Back in the classroom, some pupils will invent new sequences very readily, and will enjoy the activities. Other pupils may be slower and may need a lengthy period of short sessions to establish the concept. Most pupils will find it easier to make, and record, spatial patterns. Give slower pupils more of these to establish their confidence, before moving on to number sequences. This work is useful in helping pupils to develop concise instructions.

Express simple functions symbolically; understanding and using simple formulae or equations expressed in symbolic form.

Activity 4.19 Give the teachers practice in making up symbolic expressions.

1. Write the total cost, c pence, of n ball-point pens at 12p each.

2. Obtain formulae for the perimeter P of a rectangle with dimensions l and w cm and for a square of side s cm.
3. Use the formulae from (2) to find the perimeter of a square of side 12 cm, and of a rectangle 13 cm by 11 cm. What is the same about these shapes? How are they different?

Understanding and using terms such as 'prime', 'square', 'cube', 'square root' and 'cube root'.

During an INSET session, remind teachers that 'square' and 'square root' are inverses, as are 'cube' and 'cube root', and make sure that they know exactly what this means, before they review this material back in the classroom.

Prime numbers

The following is a useful activity for illustrating why 1 is no longer regarded as a prime number. We consider each number in turn, noticing its factors, beginning with 1.

Activity 4.20 Using centimetre-squared paper, draw as many rectangles as possible (of different shapes) for each number, so that the number represents the area of the rectangle.

- For the number 1, one possibility – it is one square.
- For 2: draw two squares next to each other horizontally. Then draw another pair of squares next to each other, arranged vertically. These two rectangles, in different positions, are to count as 2. Write the number of possible rectangles for each number underneath.
- How many different rectangles can be made for 3?
- For 4, did you find three – 2 rectangles (4 by 1 and 1 by 4) and 1 square (2 by 2)?
- The number 5 belongs to the 2-rectangle set again (5 by 1 and 1 by 5).
- How many different rectangles can be made with 6 squares?

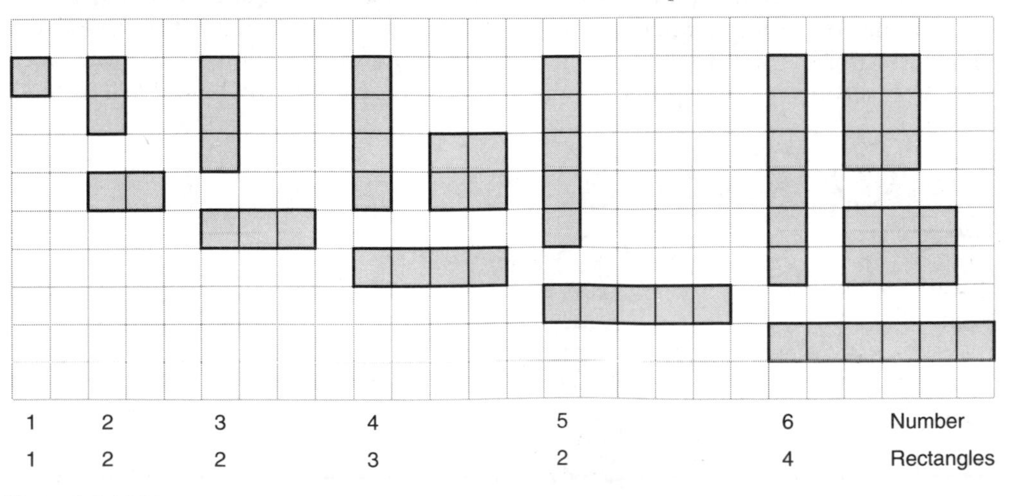

Figure 4.6 *Making rectangles*

When you have finished drawing different rectangles for every number from 1 to 20, analyse your results so far and compare them with Figure 4.6.

- Which type of number is in the set with an odd number of rectangles?
- How many of these numbers were there?
- What about the 2-rectangle numbers?
- Did you find eight of these 2-rectangle numbers? We call them prime numbers.
- How would you describe prime numbers? Do they have any factors?

Back in the classroom, pupils will need more practice than teachers need for this activity, so on the second day, ask them to investigate the numbers 21 to 30, then 31 to 40, and so on. Let them continue to 100 and display their results. They can also display other interesting facts they discover.

Characteristics of multiples To identify a number as prime, it is necessary to check that it has no factors (apart from 1 and itself).

- How can we recognize multiples of 2? (Look at the units digit.)
- How can we recognize multiples of 4? Do they know the way to determine leap year? If the last *two* digits of the year are divisible by 4, then the year is a leap year – except for the centuries! Centuries are leap years only if the previous two digits are divisible by 4; 1600 was a leap year, 1300 was not.
- Most pupils recognize multiples of 5 and 10 (last digit 0 or 5).
- Do they remember how to recognize a multiple of 9? If not, give them practice in adding the digits of multiples of 9 (to obtain a single digit).
- Also ask your pupils to investigate the characteristics of multiples of 3 and 6, and to compare these with those of multiples of 9.

With these skills they will then attack the next problem of finding prime numbers from a 6-day calendar much more efficiently.

Activity 4.21 A 6-day calendar This activity is easier if it is done on squared paper. In the first row, write the numbers 1 to 6, in the second, 7 to 12 and so on.

Continue for 30 days; when you come to a prime number, put a ring round the number. After the first row, in which columns do the prime numbers occur? Check that, after the first row, the prime numbers occur in the same two columns. Why is this?

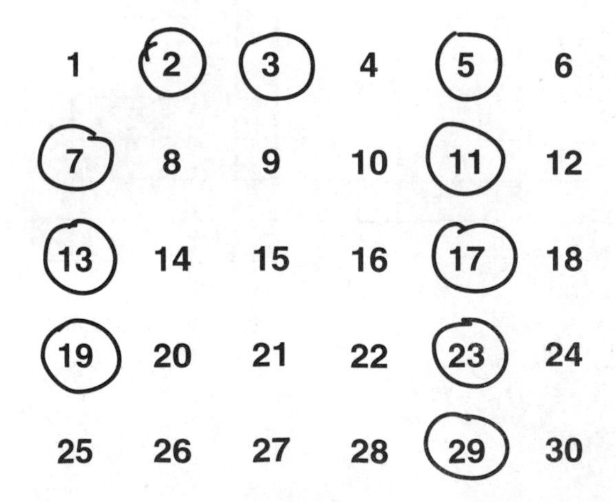

Describe the numbers in column 2, then in columns 3 and 6. Write down the numbers in column 1 which are not prime.

You may like to answer the same questions with reference to a (normal) 7-day calendar.

Coordinates in all four quadrants

> The final Programme of Study at Level 5 is:
> *understanding and using coordinates in all four quadrants*

Activity 4.22 Provide A4 centimetre squared paper for each teacher to draw two axes, intersecting at right angles, on heavy lines near the centre of the sheet. Number the axes from the origin (0,0) using a scale of 1 cm per unit, as shown in Figure 4.7.

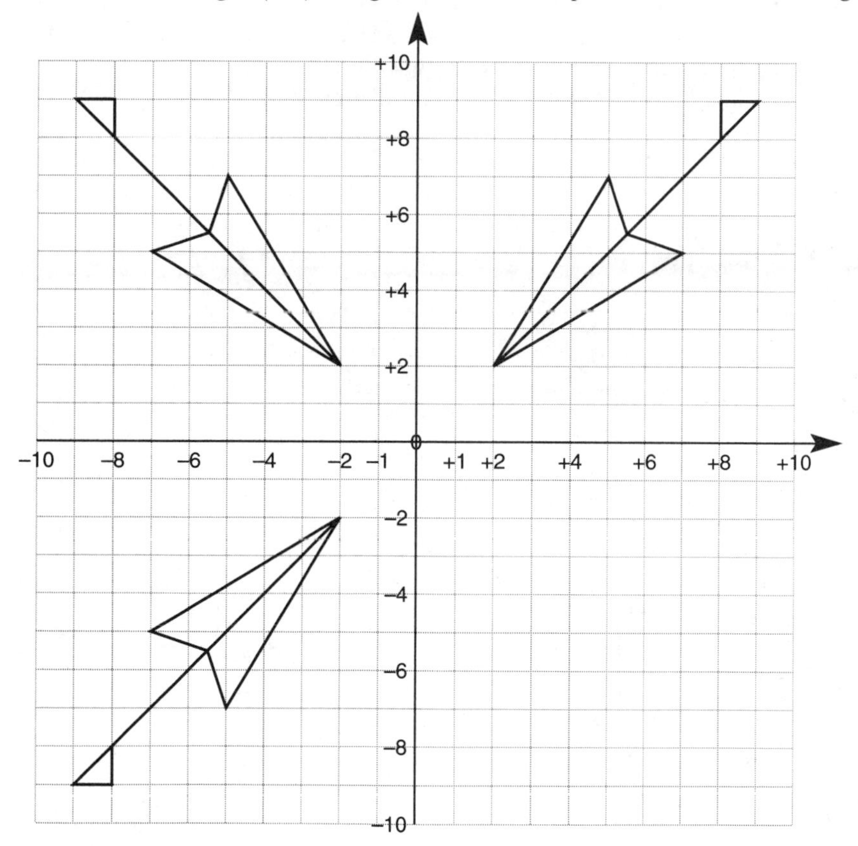

Figure 4.7 *Reflection*

1. Plot some points to draw a shape in the first quadrant. Number and list the points you have plotted, in order, together with directions for joining them.
2. Reflect the shape, using the positive y-axis as the mirror line. Number the points and make a list of them. Compare this list with the original – what do you notice?
3. Now reflect the second shape, using the negative x-axis as mirror line; list the points again.
4. Finally, reflect the third shape, using the negative y-axis as the mirror line; number the points and list them.

5. Check that the last image is a reflection of the original. Where is the final mirror line? What effect does reflection have on the coordinates?

Activity 4.23 Draw another set of positive and negative axes on squared paper as before.

1. Plot the points (0,0), (2,4), (6,4) and join them to make a triangle, and colour the triangle.
2. Then rotate the triangle about the origin, anti-clockwise through 90°. (If anyone finds this difficult, suggest that they trace the triangle and cut out their tracing. They can then pin the corresponding corner to (0,0), placing the triangle in the correct position, and then rotate it through 90°.) Draw the triangle in its second position, marking in the coordinates of the new points.
3. Then make two other rotations, anti-clockwise through 90°.

After each rotation, label the three points. What effect does rotation have on the coordinates? After the first rotation, teachers may find that they are able to calculate the new position and enter the points. Colour all the triangles (Fig. 4.8).

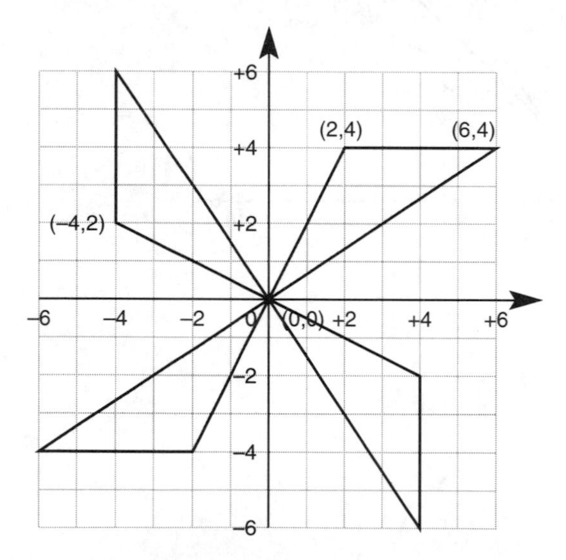

Figure 4.8 *Rotation*

Back in the classroom when trying these activities with your pupils, encourage them to make an interesting shape for Activity 4.20. They will need practice in these activities from time to time, so let them make attractive displays for the classroom.

Some children rotate shapes intuitively; others may have difficulty even in recognizing shapes made by rotation. They will need more practice in anchoring and rotating shapes and making lists of the new coordinates than in dealing with mirror reflections. Use opportunities provided by ethnic festivals (such as Diwali) for special forms of pattern-making.

LEVEL 6: GENERATING SEQUENCES, AND MAPPINGS

There are two major topics at this level:

- suggesting possible relationships for generating sequences; using spreadsheets or other computer facilities to explore number patterns; solving linear equations; solving simple polynomial equations by 'trial and improvement' methods;
- drawing and interpreting simple mappings in context, recognizing their general features.

In an INSET session review all the types of sequences generated so far; ask the teachers to suggest examples, and supplement these as necessary, to form a complete list of sequences as follows.

Addition of 3

5 8 11 14

Could you add a term at the beginning? (2)

Subtraction of 5

21 16 11 6 1

Continue for two more terms.

(−4, −9)

Difference sequences Investigate these sequences to try to find a common difference. Start with:

0 1 4 9 16 . . .

What are these numbers? (square numbers) Now calculate the first differences:

0 1 4 9 16 . . .
1 3 5 7 . . .

What are these numbers? (odd numbers) Now calculate the second differences:

0 1 4 9 16 . . .
1 3 5 7 . . .
2 2 2 . . .

The calculation of the third differences produces a row of zeros so our motto at this stage could be: In mathematics we stop at nothing!

Set the teachers to investigate cubes, and then fourth powers of numbers. How many differences are calculated before we reach a constant row? What constant is produced? Is there a pattern?

Multiplication by 5

2 10 50 250 . . .

Can they insert a term before the first?

Division by 10 Review the familiar place-value sequence:

1000 100 10 1 0.1 0.01 . . .

and make sure that the teachers' rule for generating this sequence is correct.

Ask them to describe this sequence (division by 10). Some pupils will be ready to explore these and other sequences while at the primary school, and should be encouraged to do so.

Equations

Solving problems using equations

Ask the teachers to make up some 'think of a number' problems, but to write down the equation to be solved. They can use an empty box, or x or some other letter for the unknown number. Suppose their problem is:

When I double my number and add 3, the answer is 29.

Let n be the number. Then

$$2n + 3 = 29$$

Subtracting 3 from both sides gives

$$2n \quad = 26$$

and dividing by 2 gives

$$n \quad = 13$$

(Check: LHS = 26 + 3 = 29.) What was the order of operations in the original problem? (I doubled the number, then added 3.) To solve the problem what did you do? (Subtract 3, and then divide by 2.) The order is reversed. Use all the 'think of a number' problems offered. Collect some of them to use with the pupils, as they will need plenty of practice.

Solving equations

It is very important that both teachers and pupils should understand that equations are essentially statements of the identity in value of two expressions, and therefore we must maintain the equality of the two sides: whatever operation(s) we apply to one side of the equation, we must apply in exactly the same way to the other side.

As an example we shall use a model of balance scales, using identical boxes (assuming these have no weight), each containing the same (unknown) number of identical counters, and further identical counters. Let b be the number of counters in a box. To find the number of counters in a box, we need to solve

$$4b + 1 = 3b + 5$$

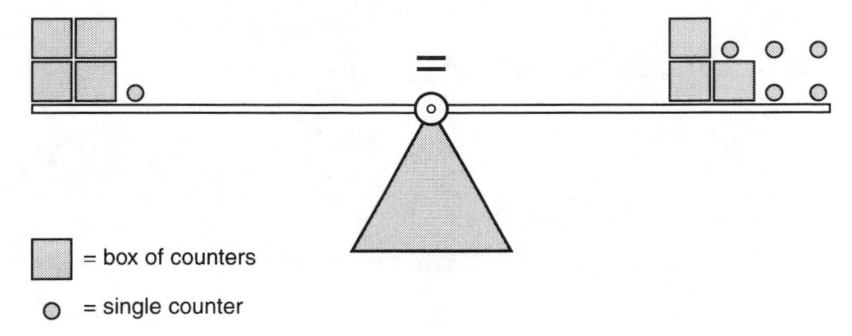

On the balance scales we would need a single box on one side and then the number of counters on the other side will reveal how many counters must be in the box. So let us remove 1 counter from each side. The balance will now look like

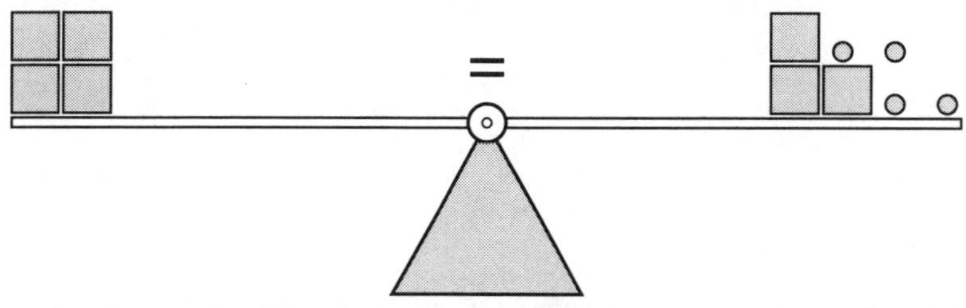

and our equation becomes

$$4b + 1 - 1 = 3b + 5 - 1$$

or

$$4b = 3b + 4$$

Now, remove 3 boxes from each side:

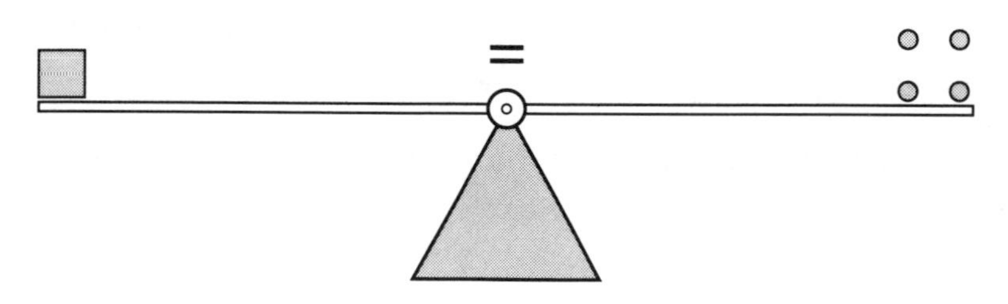

$$4b - 3b = 3b - 3b + 4$$

or

$$b = 4$$

Activity 4.24 Solve

$$3x + 4 = 12 - x$$

We imagine the balance scales and aim to collect all x terms on the left-hand side, and numbers on the right-hand side.

To leave only xs on the left-hand side we begin by subtracting 4 from both sides.

$$3x + 4 - 4 = 12 - x - 4$$

$$3x = 8 - x$$

Now to collect all xs on the left-hand side we must then add x to both sides.

$$3x + x = 8 - x + x$$
$$4x = 8$$

To find the value of x we now divide both sides by 4:

$$x = 2$$

(We can check from the original equation: LHS $= 3 \times 2 + 4 = 10$. RHS $= 12 - 2 = 10$.)

In this example, we had to 'subtract 4', 'add x' and 'divide by 4'. The first two operations could have been done in reverse order to achieve the same effect of isolating the x term. In the next example the two operations needed are: 'add 8' and 'subtract $2x$'.

	$5x - 8 = 2x + 1$		
Add 8:	$5x = 2x + 9$	Subtract $2x$:	$3x - 8 = 1$
Subtract $2x$:	$3x = 9$	Add 8:	$3x = 9$

Having isolated the x term we can then

divide by 3: $x = 3$

(Check: $15 - 8 = 7$; $6 + 1 = 7$.)

Discuss this 'balance scales' method in your INSET group. At first, teachers may find it more difficult than the method they were taught, but this new approach is important since it avoids difficult (non-mathematical!) statements such as 'when you change the side (of the equation) you change the sign'.

Solving simple polynomial equations by 'trial and improvement' methods

Linear equations can be solved by manipulating the equation until it becomes $x = \ldots$ With polynomial equations we are left with $x^2 = \ldots$ or something similar. We shall solve these equations by using a calculator.

Solving simple polynomial equations (e.g. $x^2 = 20$, $x^3 = 30$ etc.) by trial and improvement methods, using a calculator

Back in the classroom review square root and cube root, to help pupils to suggest an initial estimate. Calculators should always be available. Here are some examples which teachers will want to work themselves before giving them to their pupils.

Activity 4.25 Finding square roots. Suppose we want to know the value of x, if $x^2 = 90$. We know that $9^2 = 81$ and $10^2 = 100$ so x must be somewhere between 9 and 10. We may try 9.5 as a first estimate (half-way between 9 and 10). Now $9.5^2 = 90.25$, which is too large so x must be somewhere between 9 and 9.5 but probably closer to 9.5. We have: $x = 9$, $x^2 = 81$; $x = 9.5$, $x^2 = 90.25$.

We now try $x = 9.45$. $x^2 = 9.45^2 = 89.3025$. Which is closer to 90, 90.25 or 89.30?

So far we have established the following values

x	x^2
9	81
9.45	89.3025
9.5	90.25
10	100

We can then try 9.475, and more values for x, each time getting closer to the value 90 for x^2. We can stop as soon as the required level of accuracy has been reached. (Answer is 9.4868329 . . .)
Another solution: we know that if $x^2 = 81$, x, the square root of 81, is 9. If $x^2 = 90$, estimate x. Someone may estimate 9.5. Using a calculator, we find that $9.5^2 = 90.25$. We might decide that $x = 9.5$ is near enough. However, if we chose $x = 9.49$, $x^2 = 90.0601$. This answer is closer but still a little too large. Let's try $x = 9.489 \times 9.489 = 90.04$.

By gradually decreasing the value of x, we can get closer to 90.

Activity 4.26 A slightly harder problem is to find x if $x^3 = 20$.

Now $2^3 = 8$ and $3^3 = 27$, so x lies between 2 and 3. Try $x = 2.5$ on your calculator: $2.5^3 = 15.625$ (far too low); $2.75^3 = 20.797$ (too high – but close); $2.7^3 = 19.683$ (too low again) and so on.

When reviewing square roots and cube roots back in the classroom, give your pupils extended oral work when you want them to suggest an initial estimate; calculators should always be available. Here are some further examples which teachers will want to work themselves before giving them to the pupils.

(i) $x^2 = 10$ ($x = 3.16$ to 2 d.p.)
(ii) $x^2 = 60$ ($x = 7.75$ to 2 d.p.)
(iii) $x^3 = 60$ ($x = 3.9$ to 1 d.p. $x = 3.91$ to 2 d.p.)

Cartesian coordinates

At Level 6, the final Programme of Study is 'graphical representation using all four quadrants'. The first topic to revise with pupils is negative (and positive) numbers. Give them examples from a floor number line, if it is still there – otherwise the pupils can make a new positive and negative number line on graph paper. Give them instructions for addition and subtraction, using positive and negative numbers. Do they remember that -3 is a point on the number line? Can they find out how many steps will take them from -3 to $+5$? And from -3 to -7? Do they remember that – (-2) is the same as $+2$ and – $(+5)$ is the same as -5 because the subtraction sign has the effect of inverting the operation? Then start on multiplication. When introducing pupils to functions, base these on the number squares they have made already. Introduce straight lines before functions represented by curves. Proceed slowly, and keep the pupils' interest. Not many of them will be able to tackle this work while they are still at the primary school, but some will enjoy algebraic graphs and they should receive every encouragement.

Straight-line graphs

Looking at the straight-line graphs, the first addition table for pairs of numbers 1 to 6 (Fig. 4.1), we recall that one pattern noticed was the line of 7s along a diagonal. Along this line we have pairs of numbers whose sum is 7. This relationship can be plotted with one number being taken from the x-axis and the other number from the y-axis. Then all the points on this diagonal line belong to the relationship $y + x = 7$.

Use centimetre-squared paper to draw positive and negative x- and y-axes for values -6 to $+6$ for both x and y and draw the line $y + x = 7$ on your new axes.

If the values you choose for x are:	0	1	2	3	4	5	6	7
then corresponding values for y are:	7	6	5	4	3	2	1	0

Plot these points (Fig. 4.9). (They should all fall in the first quadrant.) Are they in a straight line? (If so, we do not need to plot all seven. When we know that we shall obtain a straight line, we can plot three points, two to fix the line, and the third as a check.)

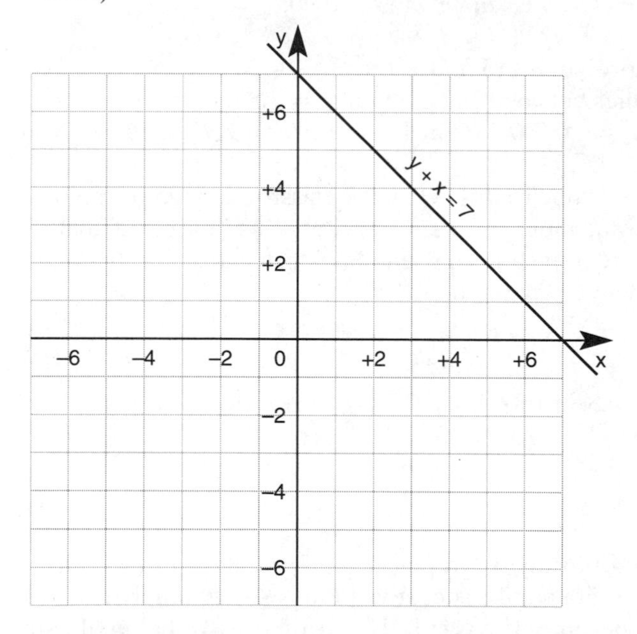

Figure 4.9 *Straight-line graph*

We shall first extend the representation of relationships, functions, or mappings, to include negative numbers. Extend the line at both ends into the second and fourth quadrants. Write down the coordinates of some points on the line $y + x = 7$ in the second and fourth quadrants, e.g. $(-1, +8)$ and $(+8, -1)$. Check that the points belong to the relationship $y + x = 7$, i.e. $-1 + 8 = 7$ in both instances. Check that all the points you have listed belong to the relationship.

Activity 4.27 On another set of axes draw the graph of $y - x = 2$ (which approximates to the line of 1s in a subtraction graph) (Fig. 4.10).

Did you expect the relationship to be represented by a straight line? Extend the line into the third quadrant and make a list of the coordinates of the new points. Do you have $(-2, 0)$ and $(-3, -1)$ on your list? Check that these coordinates belong to the relationship $y - x = 2$.

$$0 - (-2) = 0 + 2$$
$$-1 - (-3) = -1 + 3 = 2$$

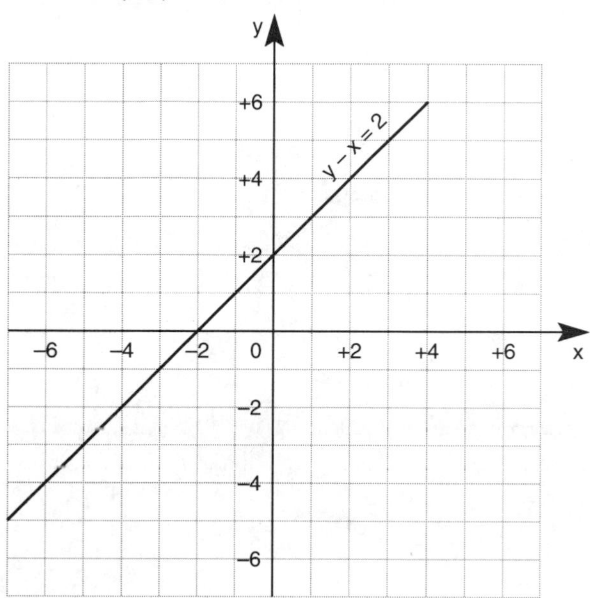

Figure 4.10 *Another straight-line graph*

Activity 4.28 Consider the columns (or rows) of the multiplication table (Fig. 4.2). These belong to the relationships $y = x$, $y = 2x$, $y = 3x$ and so on. These are all 'linear', so you may use three points to fix each line. What do you notice about the slopes of the lines?

Curved graphs

Hyperbolas In the multiplication square (Fig. 4.2) to what relationship do the points on the (curved) line of 12s belong? The products of the coordinates of these points were all 12, so $xy = 12$. Plot the four points of this function on another set of axes: (2,6), (3,4), (4,3) and (6,2). You might also include: (12,1), (10,1.2), (1.2,10); (5,2.4), (2.4,5), (1,12). Notice that when we join these points they form a curve, not a straight line (Fig. 4.11). To draw a reasonably accurate graph we need to plot many more points than three!

We now want to plot the points for which x is negative: $x = -1, -2, -3$, and so on. When x is negative, y must be negative also, since the function $xy = (+)12$, e.g. $(-1)(-12) = +12$. The third quadrant contains points with both x and y negative. How do you know that there is no other branch of this function in either the second or the fourth quadrant? Join the points to form two separate curves, one in the first and one in the third quadrant. This curve is called a hyperbola.

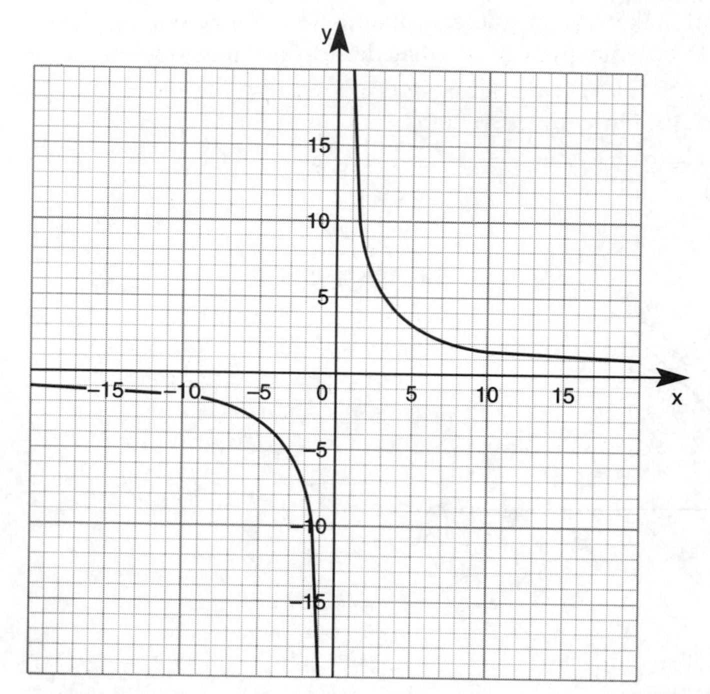

Figure 4.11 *A hyperbola:* $xy = 12$

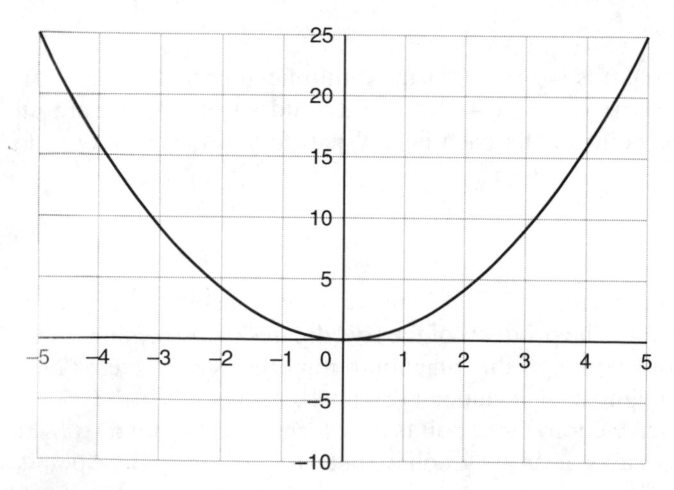

Figure 4.12 *A parabola:* $y = x^2$

Quadratic functions and parabolas Consider the function $y = x^2$. How do you know that this graph will not be a straight line? We have a point of $x - x^2$ in this case – and when $x = +3$ or when $x = -3$, x^2 (or y) $= 9$, so we can expect two values of x for each value of y. This will give us a symmetrical effect, about the y-axis. Use the same scale for x and y, say 1 cm for 1 unit, but with these data

x	0	±1	±2	±3	±4	±5	±6	±7
$y(x^2)$	0	1	4	9	16	25	36	49

very few points appear and it is difficult to draw the curve through the points since the values of y are so much greater than those of x. Instead, we choose to plot a graph of the function $y = x^2$ using a different scale for x, say 1 cm for a unit of x, and 1 cm for 5 units of y. Since there are no negative values of y it would also be sensible to limit the extent of the negative x-axis (see Fig. 4.12). You will have to approximate many of the y values if you use centimetre-squared paper. Draw the graph in pencil, lightly at first, so that you can correct it. Then you can use the graph to read off the squares of 1.5, 2.5, 3.5, etc. Then find the values of x when $y = 10, 20, 30$, and other values of your choice. The function is $y = x^2$, so $x = \sqrt{y}$, and therefore you can use this graph to find both squares (y) and square roots (x).

Chapter 5

Shape and Space

LEVEL 1: INTRODUCING SORTING AND BUILDING

Children begin to handle three-dimensional shapes and build towers of bricks before they can walk. From the beginning of their lives they are interested in the characteristics of different shapes in their environment; they can discover patterns and relationships and are immediately aware of what they have achieved. This knowledge gives them confidence and encourages them later on to persevere with other aspects of mathematics which they might have found more difficult.

All young children enjoy handling three-dimensional shapes of various types and sizes, building with them and talking about what they are doing. Moreover, at the same time they encounter many worthwhile problems which teachers can encourage them to solve. These activities provide many opportunities for discussion, and for teachers to introduce new vocabulary and to ensure that individual children use it.

Materials needed at Key Stage 1

Begin by making a collection of three-dimensional shapes. Ask the children to bring containers (boxes) of many different shapes and sizes to school. Ask them to persuade their families to open cornflakes and other packets carefully for a while, so that these can be brought to school as whole as possible – intact and without tears. Make a display of a few boxes of different shapes for the children to see:

- an empty soap powder packet (not too large);
- a spice container (cylindrical);
- a soft drink can;
- an empty matchbox;
- the centre of an empty paper-towel roll.

As each container is brought into school, ask the children to talk about them, and to suggest words to describe them, e.g. tall, flat, pointed, etc. If the children cannot think of a suitable word, teachers should suggest one, and display it – attached to the shape it describes. After a few days, the new name should be moved to 'shape words list'. Refer the children to this list frequently, asking different children each day to name the shapes in the collection. Encourage them to suggest new words and add these to the class list. Organize separate collecting boxes for different shapes, and ask for the children's help in labelling these.

Assessments

Because of the practical nature of almost all of these activities, teachers should be able to judge from the way in which the pupils carry out the tasks set, discuss the problems and describe what they are doing, whether or not they have acquired the underlying concept.

Those pupils who clearly do not understand what they are asked to do will need to work on activities which they do understand to restore their confidence. Then, later, the problems these pupils found difficult can be introduced on an easier level, so that they can concentrate on acquiring the concept. These pupils need to be given encouragement whenever possible.

Programmes of Study

At Key Stage 1 pupils should engage in activities which involve:

- sorting and classifying 2D and 3D shapes, and using words such as straight, flat, curved, round, pointed, etc.;
- building 3D solid shapes and drawing 2D shapes and describing them;
- using common words describing position, such as: on, inside, outside, above, under, behind, next to;
- giving and understanding instructions for movement along a route, in a straight line, round a circle, etc.

Sorting and classifying

In an INSET session discuss sorting activities with the teachers. Ask each teacher to bring three empty containers of different shapes, so that in total there should be a good variety to consider. Prepare a vocabulary list as the teachers carry out the activities. At this level, sorting is rudimentary. Later, when the concept of 2D faces, angle and symmetry have been introduced the sorting criteria can be more complex.

Building 3D shapes and drawing 2D views

At another INSET session, using the same materials, ask each pair of teachers to build the tallest tower possible with their six containers. The tower must stand on its own – without Sellotape. Then ask the teachers to make a freehand sketch of the completed building, showing:

- what it looks like from above;
- the elevations of the front and back.

Discuss these views, and add words to the vocabulary lists, including other words describing position, above, below, in front of, etc.

It is useful, for this activity, if the building is done on the floor. Certainly when the teachers carry out this activity with their children, they will want them to build on the floor, so that the children can see the view from above and the teachers can check that this is correct.

As an extension for the teachers they can be asked to use the same six containers to construct a one-storey building which first covers the most ground area, and then the least. As before, they should sketch the buildings from above and from the front.

In the classroom, teachers will probably want to follow this up by asking the children to sketch different 2D views of shapes in the class collection: for example, of a cube, a cuboid, a cylinder, a ball, a cone and a pyramid. Sometimes show the children a view from above and from the front, and ask if they can identify the object. Some children will find this task difficult, so teachers need to be tolerant of their early efforts. Gradually introduce the names of 2D faces of shapes: square, rectangle, circle, triangle, regular pentagon, hexagon.

Looking at 2D faces on 3D shapes

To encourage their pupils to study the 2D faces that make up 3D shapes in the class collection, teachers can ask them to cover a box (or another shape) with newspaper, and then to paint it. Make sure that crayons, paste and scissors, as well as newspaper, are available. Their presence often suggests to the children that drawing round a face on to the newspaper will give them the correct shape for each. If each face is covered separately, covering does make the box stronger, although you may want the children to find this out for themselves from experience. When all the faces are covered, the children can paint them with thick paint, perhaps using different colours for different shaped faces. They should then name the shape of the 3D object and its various 2D faces. If these models, all carefully labelled, are small enough, they can form part of a mobile.

Movement along a route

The children's first introduction to movement along a route will probably arise in a PE session, by word of mouth, when the children are following an obstacle course or a maze. At a subsequent session, instructions could be given by means of symbols on the way. These might be prepared by one half of the class for the other half and the teacher. Roles can then be reversed. If these sessions go well, the children can be asked to record each set of symbols in the class book.

Problems at Key Stage 1

Problems can arise when children are using shapes for a specific purpose, due to their stage in development. Some schools encourage the children to begin building with boxes as soon as they start school.

At the earliest stage, the children usually choose two boxes and put one on top of the other, often without any clear idea of what they are making. They then decide what it looks like – a boat, a train, or a house – and tell the teacher: 'I made a boat.' The teacher then writes a suitable label for the model.

Some months later children start by deciding what it is they want to make, and then they search for suitable boxes. They will even cut up larger boxes if they cannot find what they want. They tend to talk about the model in more detail, usually working with another child. They often cover each face of their model separately with newspaper, then paint it with thick paint. Frequently they write a description of their model, asking the teacher how to spell new words.

During the second year (at Level 2), children become more ambitious. They encounter and solve problems:

- making pitched roofs for their houses;
- attaching a chimney to a pitched roof;
- attaching wheels to a train or cart.

Attaching wheels For this, they need to have the right materials to hand: tin lids, cotton reels and empty film spools for wheels, offcuts from dowel rods of different thicknesses for axles. Their first task is to find the centre of the wheel in order to attach it to the axle. Teachers sometimes help the children by asking them to draw round the wheel on paper, cut the circle out, and find the centre. At this stage, the children may well think of folding twice to find the centre. They then put the paper circle on top of the wheel and prick through the marked centre to the wheel itself. They use a drill to make the holes for the axle (carefully supervised) and push the wheels onto it.

KEY STAGE 1: INTRODUCING ANGLES

At Level 2, the Programmes of Study involve further activities with common 3D and 2D shapes, and activities involving angles:

- recognizing shapes, rectangles, circles, triangles, hexagons, pentagons, cubes, cuboids, cylinders and spheres and describing their properties;
- recognizing right-angled corners in 2D and 3D shapes;
- recognizing types of movement – straight (translation), turning (rotation);
- understanding angles as a measurement of turn;
- understanding turning through right angles.

By now, children should already recognize cuboids, cubes, cylinders and spheres, so teachers should check that individuals can name these shapes, and can find the correct label for each. The pupils should also be able to recognize the 2D shapes given on the list above.

Tessellations

The children should know from handling these 2D shapes which of them will tessellate together exactly without leaving gaps and which will not. However, teachers should check this in their own INSET group before they try to assess individual children.

Activity 5.1 Provide teachers with sets of identical plastic shapes (Fig. 5.1): squares, rectangles, regular (equilateral) triangles, isosceles right-angled triangles, regular pentagons, hexagons, octagons. Ask them to find out which of these shapes will tessellate, and to make two lists (of shapes which will not tessellate as well as those which will). Ask them how this property can be used in the classroom, e.g. in area, when covering shapes, and in calculating the interior angles of regular shapes. Then ask the teachers to investigate these properties.

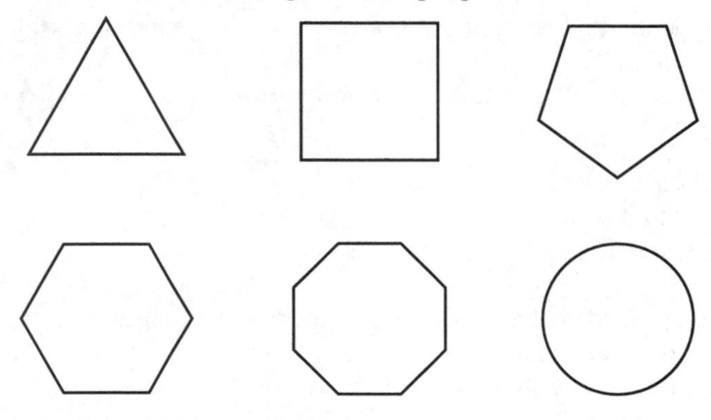

Can you name all these shapes?

Figure 5.1 *Not all regular shapes tessellate*

Assessment It is important to discuss beforehand the assessment which teachers of Key Stages 1 and 2 will make of individuals to reveal whether (or not) they can name all of the shapes listed and can describe their properties; whether they understand tessellation, and know which shapes do tessellate. Then, after the assessment of the children's understanding of the tessellation of regular shapes is complete, the results should be discussed at a subsequent INSET session. If necessary this may lead to a review of teaching methods, materials used and/or assessment procedures.

Activity 5.2 Provide groups of children with identical squares made of card or plastic, scissors and Sellotape, with which they are to make 3D shapes. Ask them to record the number of faces, edges and corners for each shape created, and to name the shape. Repeat this activity, but using different (identical) 2D shapes each time. Display the records made.

Right angles (square corners)

Square corner tester For an INSET session, ask each teacher to bring a scrap of strong paper, e.g. wrapping paper, about the area of a quarter of an A4 sheet. Ask them to fold the piece in two, pressing the fold firmly, and then to fold it again, into 4, bringing the two parts of the first fold together. They should press this second fold firmly to create a right angle (or square corner) tester (Fig. 5.2). The right angle can be coloured, back and front, so that it can be seen more easily. The teachers can then test the corners of the room, and of the walls and floor, to see which pairs are at right angles to each other. The tester can also be used to check the corners of picture frames and mirrors, and of sports fields. The teachers should make a list of their findings. Ask the teachers also to make a list of which 3D shapes include right-angled corners.

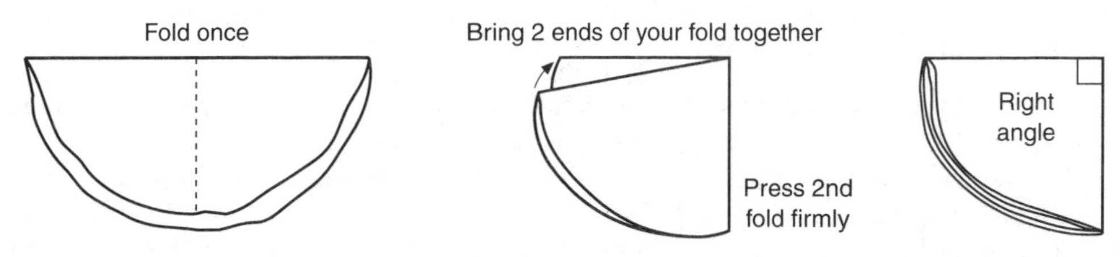

Figure 5.2 *Making a square corner*

Back in the classroom, pupils should be given short practice sessions until they become accustomed to recognizing right angles.

Activity 5.3 (This activity could also serve as an assessment.)

- Ask them to look at the angle tester when they have undone first one fold, then two folds.
- Ask them what angle two right angles make (a straight angle), then four right angles (a complete revolution).

Types of movement: translation and rotation

These two types of movement can be introduced to the children in a PE session. Observe which children confuse right and left, and give them frequent practice until they always move correctly.

Activity 5.4 Ask the children to move in a straight line until you clap your hands (just once), then to turn to the right through one right angle (a quarter turn). When you blow the whistle, they set off again in the direction in which they are facing. When you clap your hands again (this time, twice), they again turn to the right, but this time through half a turn (two right angles). At the whistle, they set off in the direction in which they are facing, until you clap again. This time you clap three times and they turn to the right through three right angles. At this point ask them how many right angles they have turned through altogether. (How many remember that this is six right angles – one and a half complete turns?) When you next clap (four claps) they make a right turn of four right angles, a complete turn.

After an interval, this activity can be repeated. This time the children turn to the left. Alternatively, you can vary the number of claps used. On another occasion, you can ask a child to give comparable directions, or ask the children to make a picture of how they moved.

Angles as a measurement of turn

Co-ordinators may find it interesting to begin an INSET session on angles with a discussion with the teachers on what they think an angle is. The answers may be surprising.

An angle is a measurement of turn.

Activity 5.5 Fasten two strips of different lengths (cut from strong card) together at one end with a push-through paper fastener. Let each group of children start with a model which they set at a zero angle.

- Ask them to rotate the shorter strip until they estimate that the strips make a right angle. (They can check their estimate, using their angle tester.)
- Ask them to continue rotating the shorter strip through another right angle, making an angle of two right angles.
- They should go on to make angles of 3 and then 4 right angles (Fig. 5.3).
- Ask them how they could make an angle of half a right angle. (Do they think of making another right angle by folding a piece of paper twice and finally folding this right angle in two?)

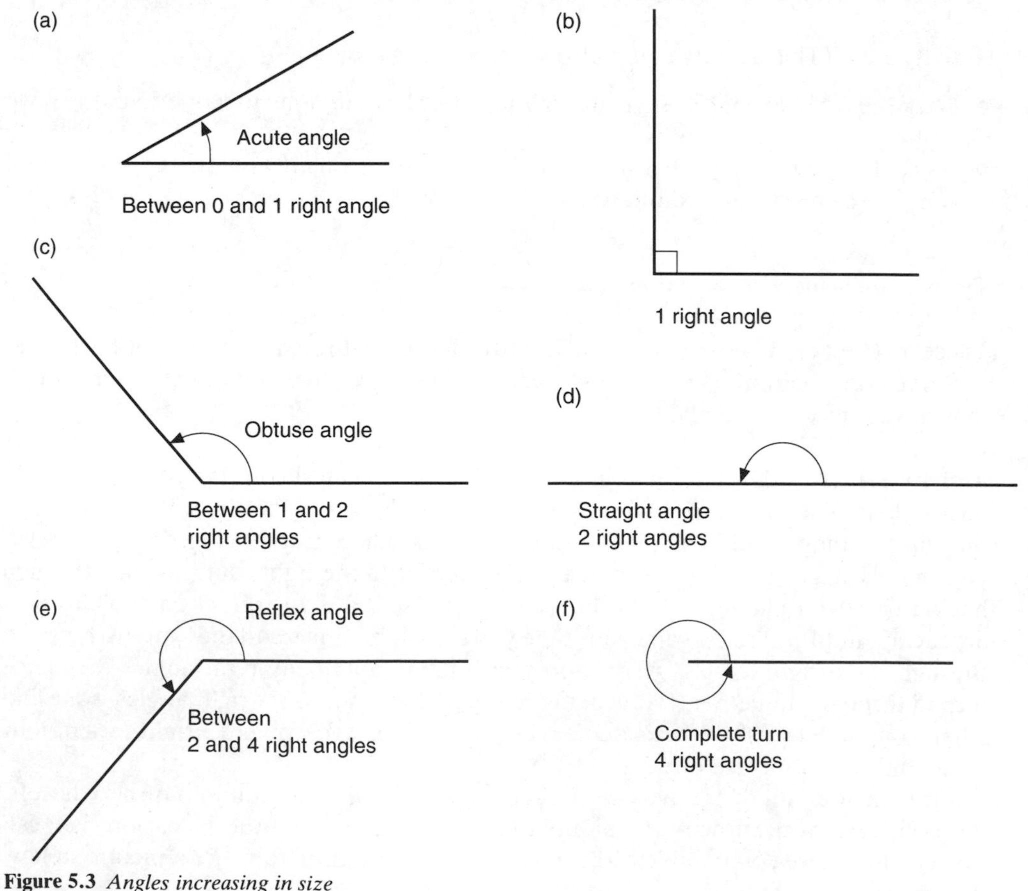

Figure 5.3 *Angles increasing in size*

Ask the children to look for angles as examples of turning. They may suggest opening a door, turning a tap, the hands of a clock. This will be the best time to introduce rotation in 'clockwise' and 'anti-clockwise' directions. Give the children instructions to follow in PE using these words. (This activity can be used as an assessment, but remember that the children will profit from frequent short practices at first.)

KEY STAGE 2: INTRODUCING SYMMETRY AND COMPASS DIRECTIONS

At this level pupils are expected to cover the following activities:

- sorting 2D and 3D shapes and giving reasons for each method of sorting;
- recognizing (reflective) symmetry in a variety of shapes in two and three dimensions;
- using and understanding 'compass directions' and the terms 'clockwise' and 'anti-clockwise'.

Sorting shapes

Activity 5.6 This activity can serve as an INSET session, or be used in the classroom. Ask the teachers to bring two identical small cuboids and two other different empty containers with them. Ask them, in groups of three or four, to sort their containers, and then focus attention on the work of each group in turn. Ask the teachers from the other groups to guess the criteria they had used for sorting and after discussion, make a list of these criteria:

- faces flat or curved
- edges straight or curved
- shapes with a square corner
- shapes with identical faces or sides, etc.

Nets

The activity which follows emphasizes the relationship between a 3D hollow shape and the 2D faces of its net (the shape made when its surface is cut so that it can be laid out flat). When using these activities with their pupils teachers should make sellotape available in case of accidents! They may also need to show their pupils how to treat sellotape so that the end remains available.

Activity 5.7 For this activity, you will need pairs of identical cuboids, newspaper, scissors, thick paint in three or four colours and big brushes. Each group takes a pair of identical cuboids, locates identical faces on one of them and paints all these in one colour. Then they locate any (unpainted) identical faces and paint all these in a different colour, and so on.

- How many colours did they need to use?

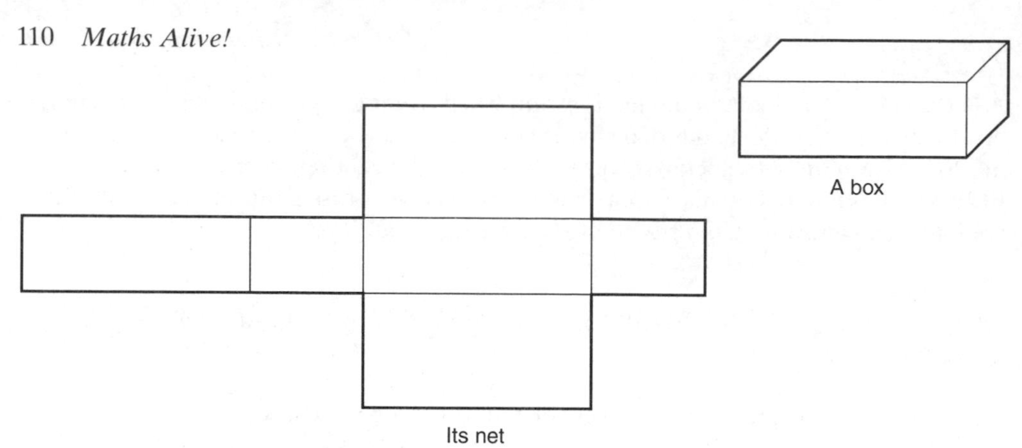

Figure 5.4 *Net of cuboid*

Then using the second cuboid in their pair the teachers cut off the flaps which help to keep the box closed, and check that the box is still intact. They then make cuts along some of the edges until the whole surface of the box (its net) can be spread out flat on the table (Fig. 5.4). They then paint the faces of the net in the same colours they used for their first box.

- How many different patterns of nets have been made?
- How many different nets could have been made?

The teachers can then make (or sketch) nets for other shapes of container, such as a cylinder, a Toblerone box, and so on.

Finally, the teachers should examine the containers (and their nets) for edges (and sides) which are of equal lengths.

- If they use the same colour for all lines of equal length on the 3D shape and its net, how many different colours are needed for each model?

Back in the classroom, pupils can construct a variety of nets, provided they can supply two identical containers. Ask them to display their painted containers together with their nets and a note describing the special features of the pair.

Reflective symmetry

We can find examples of reflective symmetry in two and three dimensions all around us. In 2D symmetry there are three pattern groups: translation, rotation and reflection.

Activity 5.8 Ask the teachers to bring patterned scarves, ties, paper (including a book of wall-paper if possible), and pieces of fabric. Provide mirrors, and the four Mirror books of Marion Walter (published by André Deutsch). Ask them to sort their patterns, according to the three pattern groups. Then ask them to find examples of 3D symmetry (with planes of symmetry instead of axes of symmetry or mirror lines). Make available a collection of 3D shapes to be examined for planes of mirror symmetry, and start with a cube. (If they do not have a cube in their own collection, they can quickly make one from squared paper.)

- How many planes of mirror symmetry has a cube?
- Are they sure that they have included all the planes of symmetry?

Give the teachers some time to find and discuss the positions and number of the planes of reflective symmetry of all the common 3D shapes.

In the classroom it is important to provide experiences of mirror symmetry in two dimensions before going on to three-dimensional shapes. Ask teachers of younger children to make a display of the work they have covered, probably in Art or PE sessions, and to share their ideas. This will probably include:

- children's discoveries made from their observations of their actions in a long mirror (walking towards and away from the mirror, lifting the right arm or left leg, etc.);
- mirror patterns made with paint between folded paper, and by cutting folded paper;
- studying patterns of different types from fabric and paper, and sorting out those with mirror patterns.

Activity 5.9 This activity and the following one could be used as an assessment of an individual pupil's understanding of mirror symmetry. As usual, teachers should try the activity themselves before using it in the classroom or as an assessment.

Begin by asking the children to create their own shapes with mirror symmetry on squared paper. Explain that they may use two or three coloured pencils to fill in their shape. Identical shapes should be in the same colour on both sides of the mirror line. They should first mark in the mirror line, then colour whole squares, and finally describe the finished shape. These should then be checked for mirror symmetry by the teacher.

Activity 5.10 This is a game for two pupils to play: A and B. They need a rectangular peg board, with up to 16 holes across and 20 holes vertically, and pegs to fit of two colours. A line should be painted in the middle of the 'vertical' columns (between the 10th and 11th pegs) or the mirror line marked with a long elastic band. Each player selects pegs of one colour, and plays on different sides of the mirror line.

Player A puts in a peg in the top line on his side. Player B puts her peg in her own side of the mirror line so that it is the reflection of A's peg. Player B then puts a peg in the second row of his side for A to mark the reflection on her side. Then A puts in a peg for B to reflect. They continue in this way until the pattern is complete (they have reached the bottom of the board). If the teachers find the pattern to be correct, the two pupils understand mirror symmetry. If one pupil has made a mistake and is able to correct this, he or she too understands mirror symmetry – in two dimensions. Three-D symmetry can then be assessed by using 3D shapes.

Compass bearings

The co-ordinator should first find out the direction of north within the school. The INSET session should then take place in a room in which the direction N can be clearly marked – probably on the ceiling. When the teachers first come into the room, ask them to point to the North. (It is unlikely that they'll be carrying a compass, and most will work it out from the position of the sun.) Suggest that all the teachers who will be working on compass bearings with their pupils should find the North direction in their classrooms and, if the head agrees, arrange for an N to be painted on the ceiling.

Activity 5.11 This first activity concentrates on the four points of the compass: N, E, S and W. Ask the children first to face N, to turn through one right angle in a clockwise direction, and then to say in which direction they are now facing. Repeat this several times, moving through one right angle at a time but using different starting points. Later, repeat this activity, this time turning in an anti-clockwise direction. Sometimes, vary the number of right angles to be turned through. It is very important that the children should be absolutely sure of the compass directions in relation to the school and also in relation to their home. Find out if they know in which direction their front door faces.

It is also useful to have the points of the compass painted on the playground. Take your pupils into the playground. Let them use magnetic compasses to check the points of the compass painted on the playground. Ask two pupils to stand at the centre of this compass and to use the magnetic compass to find the direction, from that centre, of local landmarks. The directions of these landmarks – churches, chimney stack, town hall, clock tower, etc., should be carefully recorded.

In the National Curriculum, this topic has now been transferred to Geography. Some teachers may like to co-operate with teachers of geography or to reinforce work on compass bearings by including the following examples.

Activity 5.12 We now extend the four points of the compass to eight (Fig. 5.5). Make the largest square possible from an A4 sheet of paper, then cut it out and fold it into four equal parts through the centre. Fold again to make eight sections, adding NE, SE, SW and NW.

Activity 5.13 Locate a wind vane in the locality and take your pupils to look at it. If it is not in a busy thoroughfare, ask pairs of pupils to read the wind direction every day for a month. This can be combined with a project on the weather (Chapter 7). Make sure that every pupil knows that the arrow points to the direction from which the wind comes.

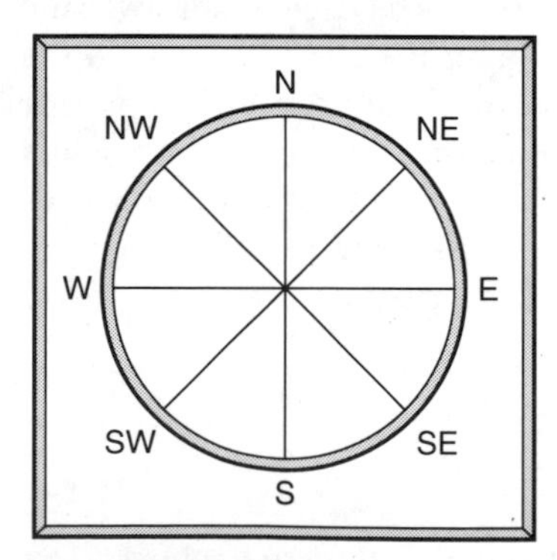

Figure 5.5 *Points of a mariner's compass*

Encourage pupils to consider the angle between wind directions when there is a change of wind. Using a magnetic compass, teachers can begin to introduce the vocabulary associated with the size of angles:

acute (sharp angle) *obtuse* (meaning blunt angle) *reflex* angle

Encourage pupils to use these words; see Figure 5.3.

KEY STAGE 2 (LEVEL 4): CONSTRUCTION AND ROTATIONAL SYMMETRY

At Level 4 the pupils will be:

- constructing simple 2D and 3D shapes from given information and knowing associated language;
- reflecting simple shapes in a mirror line – understanding congruence of simple shapes;
- understanding and using language associated with angles;
- specifying location by means of coordinates in the first quadrant and by means of angle and distance;
- recognizing rotational symmetry.

Constructing simple shapes

Co-ordinators need to make sure that the teachers involved at this level know the essential properties of the shapes listed.

Rectangles and squares There is more than one way of constructing a rectangle or a square. When you ask the pupils to say what a rectangle is, do they answer that 'its two pairs of opposite sides are equal and that its angles are all right angles'? This is not the mathematical definition of a rectangle but it *is* what they need to know in order to construct a rectangle.

Circles When teachers are about to introduce the construction of a circle, they should take pupils outside to a netball goalpost and ask them how the goal circles are drawn. The teacher should have lengths of string and sticks of chalk available (but not obviously). Could they draw a smaller goal circle? If they do not suggest tying a length of string to the post and a piece of chalk to the other end of the string, show them the string and see what happens.

When the pupils are back in their classrooms, teachers should suggest that they use compasses to draw circles of radius 5 cm, 10 cm, etc. By now they should have a basic understanding of what a circle is.

3D shapes Do all the teachers know the difference between prisms and pyramids (Fig. 5.6)? What about a cylinder and a cone? At an INSET session the co-ordinator could have some models on display, and ask the teachers what the differences are. If you have some prisms made with dog biscuits of identical shapes piled one on top of another, they will realize that prisms have constant cross-sections – they are built of equal slices.

A triangular pyramid of hollow type (tetrahedron) can be made from four identical equilateral triangles (Fig. 5.6(c)), while a square pyramid is made from a square base and four identical isosceles triangles, one on each side of the square (Fig. 5.6(b)). The teachers should make rough sketches of the nets of these two pyramids before they embark on accurate constructions. Back in their classrooms, they should encourage pupils to take the responsibility for creating their own nets for pyramids and prisms. All the pupils could then make a display of their models.

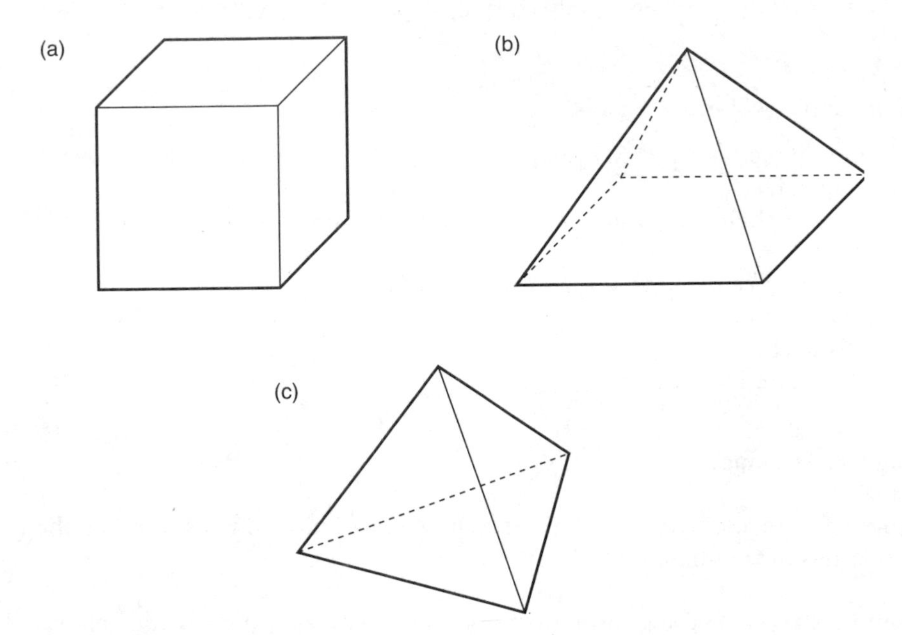

Figure 5.6 *Three-dimensional shapes: (a) square prism (cube); (b) square pyramid; and (c) regular tetrahedron*

The pupils are expected to use the correct vocabulary in the appropriate context (and be assessed accordingly):

perpendicular vertical horizontal
acute obtuse reflex

Rotational symmetry

We now consider shapes with rotational symmetry, with mirror symmetry, and with congruence.

At Key Stage 1 (Level 2), teachers may have given children boxes with lids, each of a different shape, and asked the children to find in how many ways the lid could be fitted on the box (Fig. 5.7). (Matching marks were made on the lid and box and children were usually left to discover this clue.) In order to solve this problem they had to rotate the lids. They recorded the number of positions of each lid during a complete revolution. This activity provided an early experience of rotational symmetry.

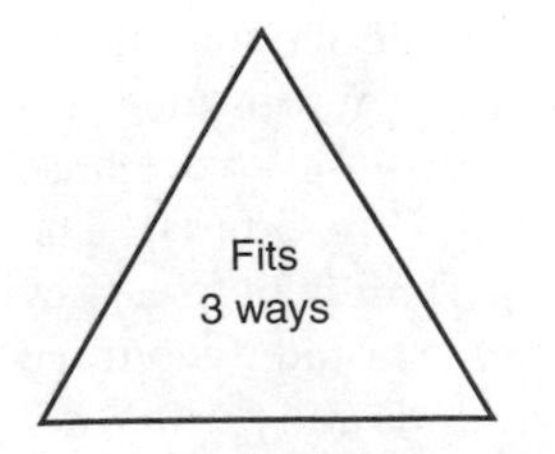

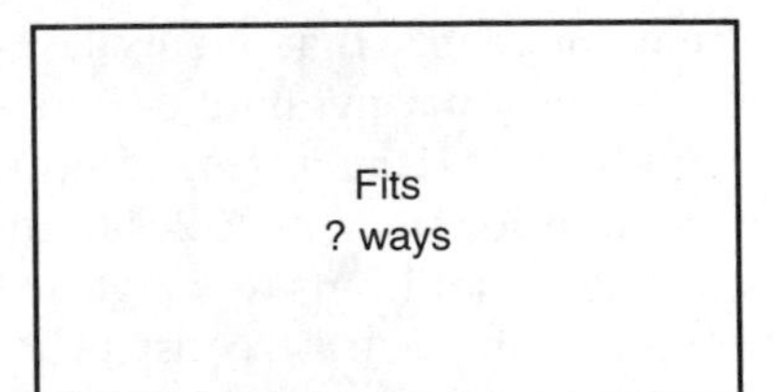

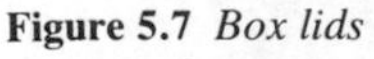
Figure 5.7 *Box lids*

Activity 5.14 Provide each person with five sheets of A5 paper (or cut half-sheets of A4 paper). Ask them to make a single fold in each of 5 sheets to divide the sheet in half, each fold to be different.

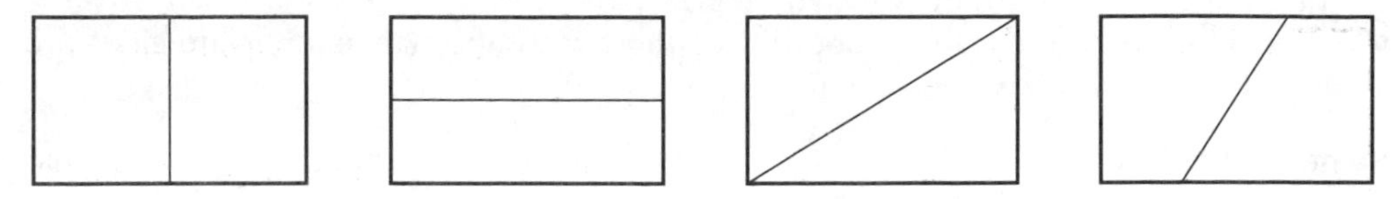

Figure 5.8 *Halving a rectangle*

Folding the first few sheets should be easy. They will begin by matching opposite sides, then folding along the diagonal, and perhaps another by matching opposite corners – which gives an oblique cut. Having exhausted the obvious ways, some may then make another oblique fold which passes through the centre by matching overlapping lengths. (If so, ask them to check this fold by holding the paper up to the light and making a second fold.) When they have checked that the shapes are identical, ask them where the two folds intersect (at the centre of the rectangle).

You can now ask them if this discovery suggests other possible folds, and how many folds there are altogether. (An infinite number!)

Ask them to take the A5 sheet with the diagonal fold, to cut along the fold and to replace the halves as a rectangle. Ask them to verify that the two pieces are identical. (Some of them will probably begin by flipping one piece about the diagonal, but they will find it does not coincide with the other piece, so the two triangles do not have mirror symmetry.)

As they continue their efforts, some will rotate one piece and will succeed in fitting it on top of the other. Ask which point was the centre of rotation (the centre of the rectangle), and what was the size of the angle of rotation (half a turn). Some will find this easier to answer if they put a pin as close to the centre as possible, and turn the pinned shape until it coincides with the other one. Or they may check by tracing one triangle and turning the tracing until it coincides with the other. These two triangles, formed by the diagonal, have rotational symmetry.

Next they should match the two right-angled triangles and colour pairs of sides of equal lengths in the same colour, front and back. Then colour equal angles, again on the front and back of the triangles, using the same colours as were used for the side opposite each angle.

Use these two triangles (called congruent triangles because they are identical) to make shapes with matching sides. If the teachers (or children) work in groups of six, they will be able to make – and preserve – six different quadrilaterals (four-sided figures) as shown in Figure 5.9. Ask them to identify the shapes with mirror symmetry, and to name these three shapes.

The two new triangles formed are isosceles, and their heights are the mirror lines (isosceles is Greek for equal legs). Ask how they know that the base of the isosceles triangle is a straight line. (The two angles at D are right angles.) The other shape with mirror symmetry is called the kite.

Now, looking at the other three shapes, do they recognize these shapes?

- Did they test them for mirror symmetry?
- If so, how?

Of course, two of the shapes have rotational symmetry – not mirror symmetry.

- Ask them to describe the centre of rotation, and the angle of rotation (the centre of the shapes; half a turn or two right angles).
- Ask them to describe the properties of the two shapes (their opposite sides are equal in length, their opposite angles are equal).

Some may remember that both pairs of opposite sides of these shapes are parallel, and that they are called parallelograms.

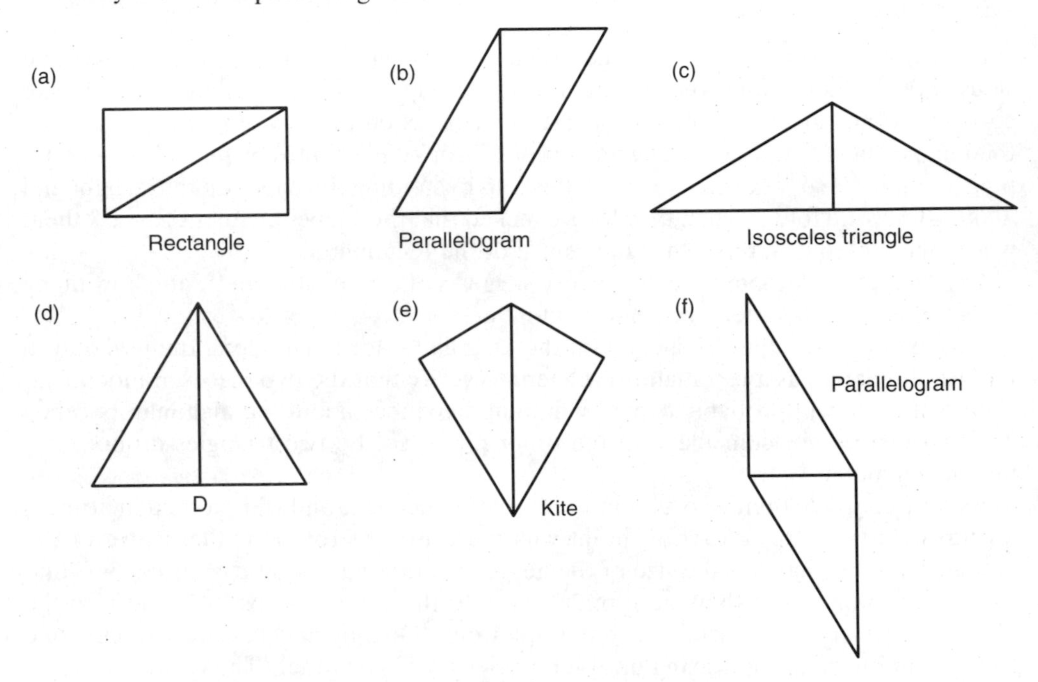

Figure 5.9 *Shapes made from two identical triangles. Which shapes have mirror symmetry?*

Teachers will recognize the sixth shape as the original rectangle! They discovered that the diagonal divided the rectangle into a pair of triangles (right-angled) with rotational symmetry.

To summarize the properties of the two right-angled triangles made by a diagonal of a rectangle:

- they have rotational symmetry; and
- they can be used to make three different shapes with mirror symmetry (two isosceles triangles and one kite) and three shapes with rotational symmetry (two parallelograms and one rectangle).

Ask:

- What is the same about all the shapes?
- Which pairs of shapes have the same perimeter? (There are 3 pairs.)
- What do you notice about the pairs with the same perimeter?

Activity 5.15 This activity offers more experience of this topic.

Take another pair of identical shapes made by folding the rectangle in half. Begin with an oblique fold which divides it into two congruent quadrilaterals (four-sided figures).

- Estimate how many different shapes you can make this time, and how many will have mirror symmetry and how many rotational symmetry?
- Do you recognize all the shapes made? Name them and describe them.
- Find the two shapes with the longest perimeter and the two with the shortest perimeter.
- Repeat the activity with the pair of shapes made by matching opposite sides of the rectangle.

Pupils will certainly need to undertake most of the activities described so far, if they are to acquire the concepts of mirror and rotational symmetry, and if they are to understand the congruence of simple shapes. The next activity will give pupils experience of all three concepts.

Activity 5.16 Each pupil needs two cardboard centres from paper towel rolls. They should work in groups of three or four. First let them inspect a towel centre, then sketch what they think the unravelled centre will look like. Soak the centres in water until they unravel. Then hang them over a string line to dry. Give each pupil four pieces of card, from two centres. First ask them to find out whether the pairs are identical (or congruent). Ask them to find out, by matching sides, which lines are equal in each pair. They should mark all equal sides in the same colour, on the back and front of each shape. Then they match equal angles and mark these. Ask them if they recognize the shapes. (They are parallelograms.) Once more, ask them to recount their properties (and use this as an assessment).

Activity 5.17 Ask the pupils to bring to school for display a collection of scarves, ties, pieces of fabric, wrapping paper and wallpaper, decorated boxes and pieces of pottery. The pupils should take a major part in selecting suitable items for display, ensuring that there are examples of traditional crafts. It is also important to include examples from as many different cultures as possible. As the pupils organize the collection, check that they include repeating patterns made by translation, by reflection and rotation. Invite parents to help; in particular ask for a parent volunteer to photograph the collection, together with those who contributed to the preparation.

Specifying location

This topic has already been partially covered in Programme of Study 2 (Algebra), at Key Stage 2 (Level 4), in the introduction to the use of coordinates for determining position.

Activity 5.18 Provide Ordnance Survey maps of the locality and give the grid reference of the school, town hall and local church.

- Can the pupils find these buildings on the map without difficulty?
- Can they also give the grid reference of three buildings of their choice, e.g. the nearest railway station?

If these tasks are completed without difficulty, discuss the basic requirements of the coordinate system: distances are recorded from two axes at right angles. However, it is not always possible to measure such distances. Ask the teachers (or pupils) to suggest conditions in which it would not be possible to measure the distance of, say, a church spire from two roads at right angles to each other. (There might be a river between the church and one of the roads, or another obstacle such as a housing estate.) Ask how the position of an object – perhaps 'buried treasure' – might be determined, say by measuring one distance and a direction(or bearing).

Bearings

Today, all bearings are calculated by assuming that you start facing North and turn, moving in a clockwise direction. So 125° means N 125° E.

- What would 200° mean? (S 20° W)
- What would 300° mean? (W 30° N)

For the buried treasure activity given next the teachers will need to sketch the directions before trying them out in the playground or school hall. Provide mariners' compasses and surveyors' tapes.

Activity 5.19 A sea captain buried a hoard of gold ducats on an uninhabited Pacific island. On his death bed he gave his eldest son directions (Fig. 5.10) for finding the treasure.

> From the lone coconut palm on the North East peninsula, walk 178 yards on bearing of 125° and dig down.

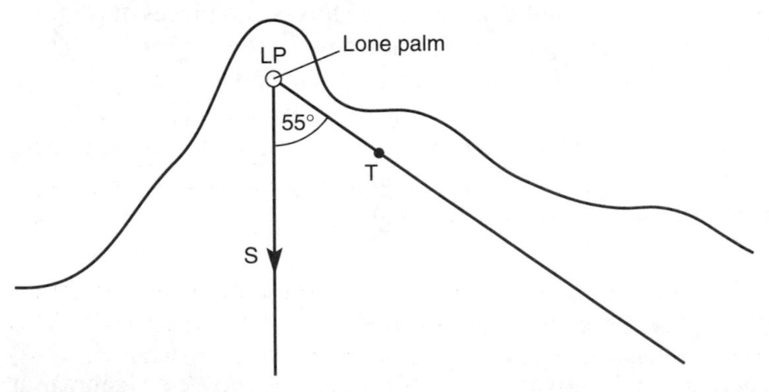

Figure 5.10

When teachers try this activity with their pupils they may choose to make it part of a more comprehensive project on surveying or it could form part of a geography project.

KEY STAGE 3 (LEVEL 5): ANGLES, VOLUMES AND AREAS

At Level 5, the pupils will be:

- identifying the symmetries of various shapes, as introduced in Key Stage 2;
- specifying location by means of coordinates in four quadrants as covered in the Programmes of Study on Algebra, Key Stage 3 (Level 5);
- measuring and drawing angles to the nearest degree;
- explaining and using properties associated with intersecting and parallel lines and triangles and knowing associated language;
- finding the circumference of circles, practically, introducing the ratio π;
- finding areas of plane figures (including circles) using appropriate formulae; and
- finding volumes of regular shapes (including cylinders) using appropriate formulae.

Intersection of two straight lines

If two straight lines are drawn (in the same plane), either they will cross or they will not. Parallel lines do not cross and are considered soon. Here we consider those that do cross. The meeting point of two straight lines is called a vertex. We need to know certain rules about the angles the intersection forms, but first we consider the use of protractors.

Measuring and drawing angles to the nearest degree

When we need to measure angles more accurately than in right angles and half right angles, we use a much smaller unit, called a degree. One right angle is divided into 90 degrees, written 90°.

- How many degrees are there in a straight angle?
- How many in a complete turn?

When we want to draw an angle of a specific size, or to measure an angle, we use a protractor. (Those bought commercially are usually made of clear plastic, and the base is extended downwards as a protection for the zero line.) Angles can be measured from 0 to 180 degrees.

Before measuring an angle decide whether it is acute (between 0° and 90°) or obtuse (between 90° and 180°) – or occasionally reflex (between 180° and 360°). This knowledge will help pupils to read the correct figure on the protractor (which has two identical scales, one running from 0° to 180°, the other from 180° to 0°).

Place the centre of the protractor at one corner of the triangle, with one side of the triangle along the base of the protractor, as shown in Figure 5.11. Then read off the size of the angle at the point where side AB cuts the protractor. Record this reading within angle A, and repeat this exercise for the other angles.

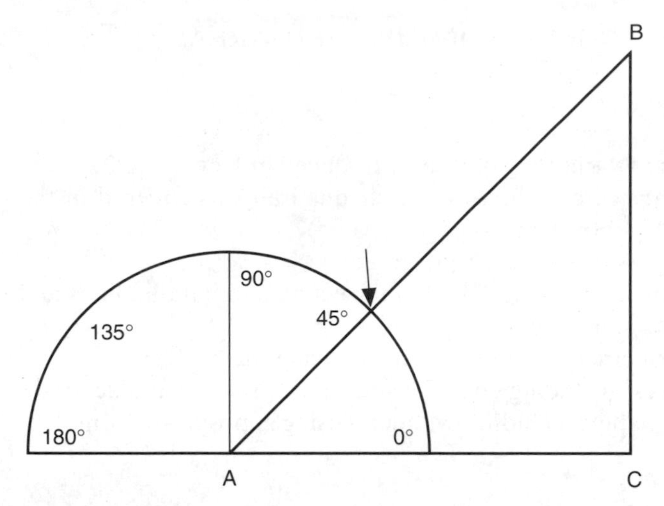

Figure 5.11 *Measuring angles*

Activity 5.20 Find the unknown angles in Figure 5.12. In Figure 5.12(b), angles *a* and *c* (and *b* and *d*) are called vertically opposite, and vertically opposite angles are equal, so $a = c$ and $b = d$.

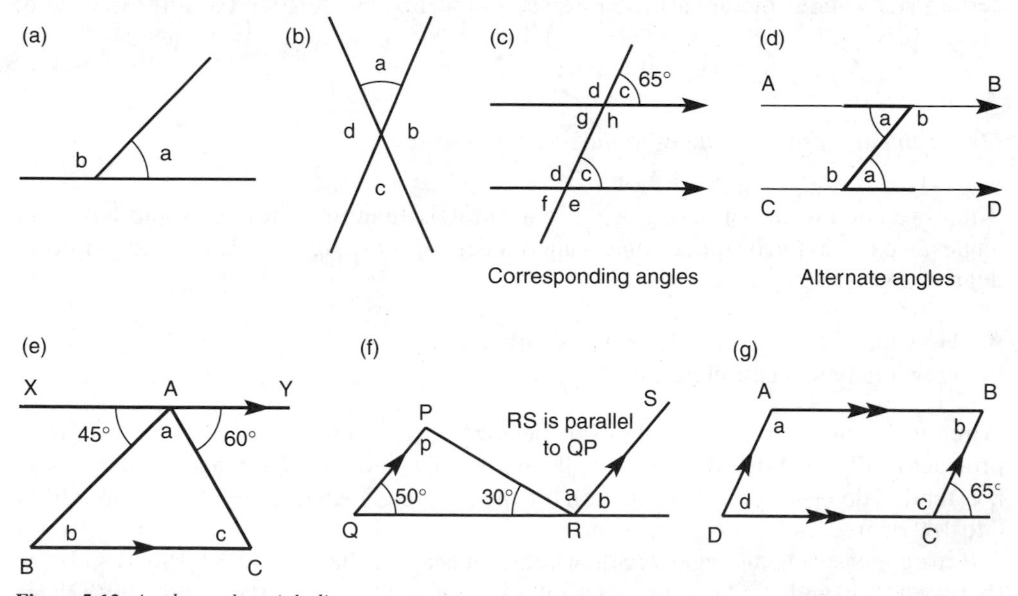

Figure 5.12 *Angles and straight lines*

Parallel lines

Parallel lines do not intersect. They are in the same plane; you could draw two 'skew' lines which never meet because they are not in the same plane, either. Your pupils may not know that we call lines such as railway lines and vertical lines on radiators parallel lines. Examples are shown in Figure 5.12 (c–g).

Your pupils may well have noticed other sets of parallel lines travelling in the same direction but never meeting. Ask them to make sketches or notes during the coming week of examples of parallel lines they see.

To draw a set of parallel lines we normally use a set square and a ruler. The ruler is used to fix the direction of the set square, which we can slide up or down, drawing lines as we go.

The lines are parallel because we have fixed the direction of the set square by keeping the ruler steady. (If pupils do not have a set square they can use a pencil box instead.) We put arrows on parallel lines, one arrowhead on the first set, two arrowheads on a second set and so on (see Fig. 5.12(g)).

Activity 5.21 Ask the pupils to draw a set of parallel lines and to draw in another line intersecting the parallel lines, as shown in Figure 5.12(c). Ask them to measure the angles *c*. Because we fixed the direction of the set square or pencil box, the angles marked *c* in Figure 5.12(c) are equal. Explain that these angles are called **corresponding angles**.

Considering the upper intersection point, angles *c* and *g* are vertically opposite and are therefore equal. In Figure 5.12(d) these angles (now marked as *a*) are called **alternate angles** and they are equal. They are on alternate sides of the cutting line. Similarly the angles marked *b* are also equal alternate angles. Pupils can recognize alternate angles because they are the equal angles within a letter N or a letter Z. Frequent short practices are necessary for sound learning.

A second intersecting line (not parallel to the first) creates a triangle between the parallel lines. The fact that angles on a straight line add up to 180° should be remembered when solving these problems. See Figures 5.12(e) and (f).

Activity 5.22 Ask them to draw two pairs of parallel lines cutting (or intersecting) each other, marking the first pair with single arrow heads and the second pair with double arrow heads. Find out whether they know the name of the four-sided shape they have made (parallelogram, rectangle or rhombus – depending on their choice of intersecting angle and position). Ask them to work out the sizes of all the angles in Figure 5.12(g). As they do this ask them to give reasons for their answers. e.g. $C = 115°$ (adjacent to 65° on a straight line). They will find that all the interior angles are either 65° or 115°. Also ask them what they notice about the opposite angles of parallelograms (and rectangles and rhombuses) – they are equal.

Accurate drawings of rectangles and triangles

Discuss the properties of rectangles with pupils. Then ask them for methods of construction.

Rectangles To draw a rectangle (Fig. 5.13) with edges 7 cm and 3.5 cm, pupils will need a ruler and a protractor. They can begin by drawing a base line, 7 cm long. At each end of this line, AB, they measure an angle of 90°, using a protractor.

They make these lines, AD and BC, 3.5 cm long. Ask them to join DC and to measure its length. It should be 7 cm long, and the angles ADC and DCB should be right angles.

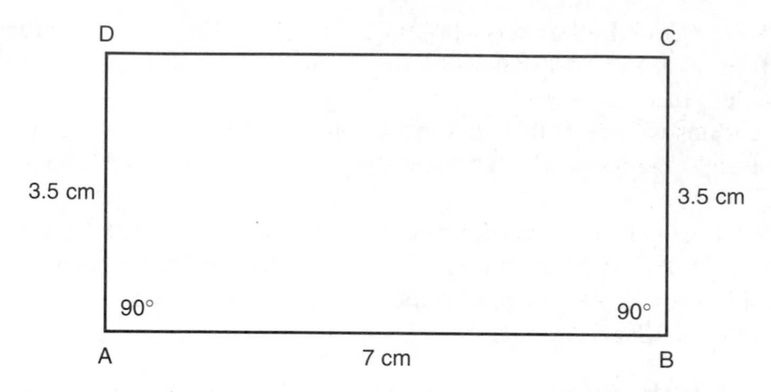

Figure 5.13

Triangles Triangles may be specified by one side and two angles (Fig. 5.14(a)) or two sides and the angle enclosed (Fig. 5.14(b)) or the length of all three sides. Only the first two can be drawn using only a protractor; the third requires a compass!

Activity 5.23 Refer to Figure 5.14(a). Draw the base 7 cm long (AB). At A, use your protractor to make an angle of 36° with AB, and draw a line AC and beyond. At B, make an angle of 55° with BA. Draw BC so that AC and BC meet at C. Measure the angle at C. Also measure the lengths AC and BC.

Now refer to Fig. 5.14(b). Draw AB 6 cm long. At A draw an angle of 25° with AB, and make this line at least 4 cm long. Mark a point C exactly 4 cm from A. Measure BC.

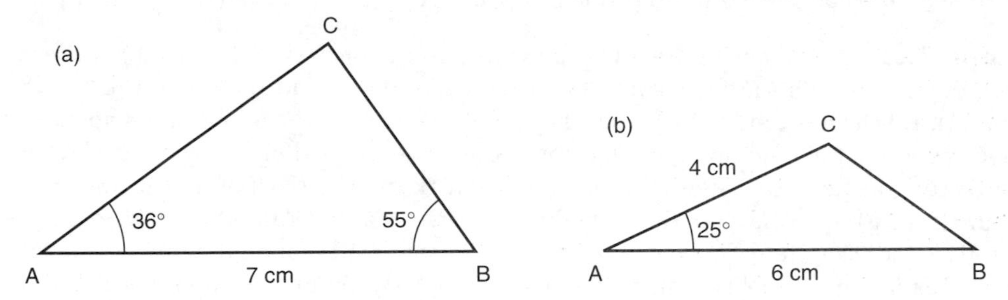

Figure 5.14 *Two methods of constructing triangles*

Adventures with angles: the sum of the angles of any triangle. Some surprising results

Do you remember a surprising fact you learned at school: that the sum of the angles of a triangle, whatever its shape, is always the same? Draw a triangle, mark its angles, tear them off, and put them close together, one after the other, as in Figure 5.15. You should find that the outer lines make a straight line, and so the angle is two right angles; the 'straight' angle made by the three angles of a triangle is 180°.

Test the angles of triangles of various shapes (right angled, acute and obtuse) to convince yourself that the sum of the angles is 180°. You may like to try this with your pupils and then give them the following activities as practice.

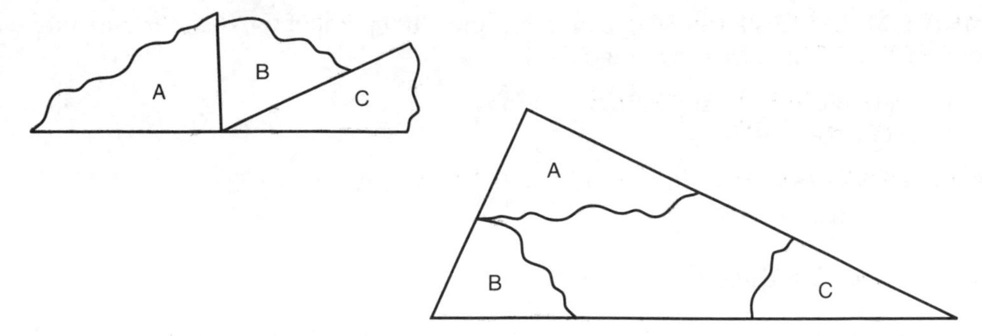

Figure 5.15 *Finding the sum of the angles of a triangle*

Activity 5.24 Find the angles of these triangles:

1. a regular (or equilateral) triangle (having equal sides and angles);
2. a right-angled triangle with two equal angles;
3. an isosceles triangle (having two equal angles) with one larger angle of 100°;
4. an isosceles triangle with two equal angles of 30°.

Activity 5.25 The outer shape in Figure 5.16 is a rectangle, and $b = 2a$. Find the angles a, b and c.

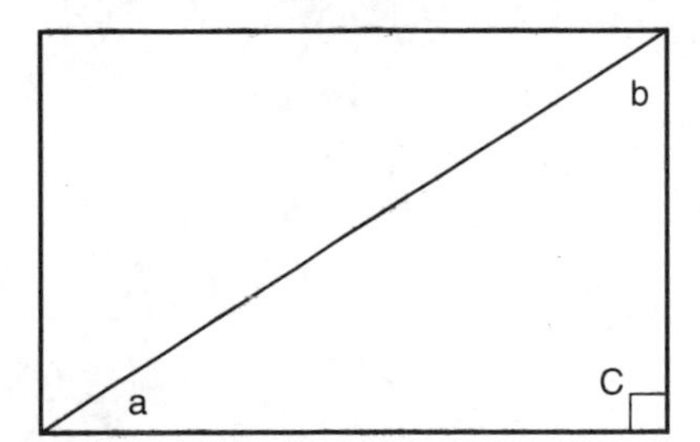

Figure 5.16

Activity 5.26 The outer shape in Figure 5.17 is a regular pentagon (equal edges and angles). How many triangles are there? What is the sum of the angles of three triangles? So what are the equal interior angles of this regular pentagon?

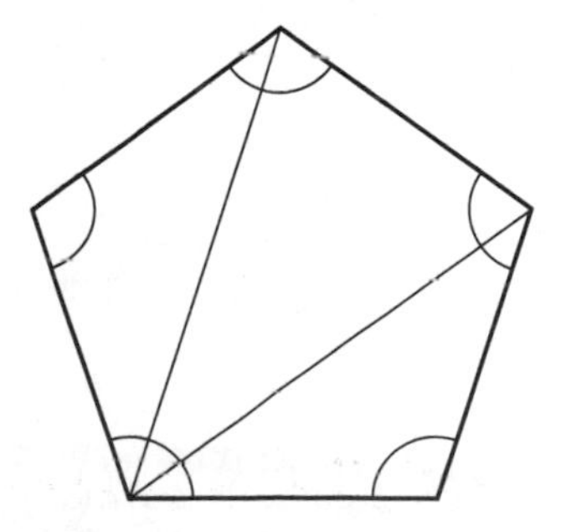

Figure 5.17 *A regular pentagon*

Activity 5.27 ABC is a triangle. Through B draw PQ parallel to AC. Mark the three angles of △ ABC as a, b and c. Find

1. an angle alternate to angle PBA = a (alt)
2. another angle QBC = c

What does this show about the sum of angles PBA, ABC and CBQ?

Representations using angles: pie charts

Once children can guess the approximate size of angles in half or whole right angles, and later on, in degrees, they can be asked to interpret pie charts. In these circular charts three or four pieces of information are represented by 'a piece of pie'. The size of each section is shown by the angle made at the centre of the circle. For example, Figure 5.18(a) shows the way a young baby spends his day: 18 hours (3/4 of day) sleeping, 1/8 feeding, 1/8 crying or playing – altogether 1/4 (6 hours) awake. Figure 5.18(b) shows a toddler's day.

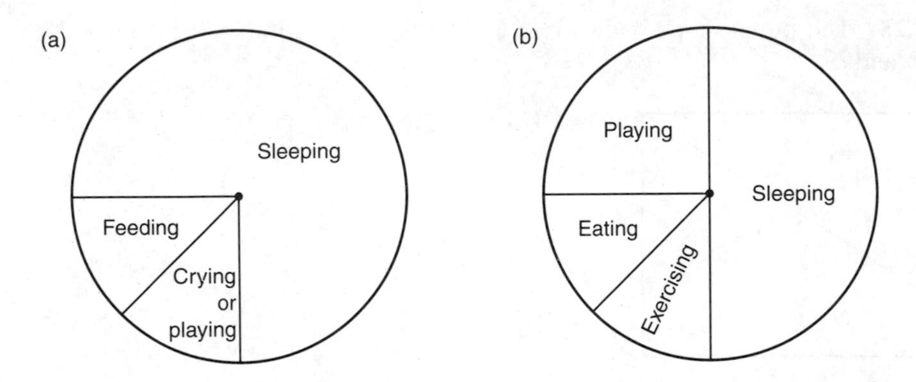

Figure 5.18 *(a) A baby's day; (b) a toddler's day*

Activity 5.28 Ask your children how the toddler spends most time and least time during the day. Then help your children to make a simple pie chart to show how they spend a day at the weekend.

Volumes and areas

Cuboids

The volume of a solid prism is:

area of base × the height

Provide centimetre cubes for the pupils to investigate the volumes of cuboids and other prisms.

Activity 5.29 In groups of four, ask them to make as many cuboids of different shapes as possible, all of volume 36 cubic centimetres.

- Record, for each cuboid, the dimensions and the volume.
- Record, for each cuboid, the product of the area of the base and the height.
- Verify that the volume of all the prisms can be found by multiplying the number of cubes used to cover the base (its area) by the number of cubes in the height.

Encourage the pupils to make prisms of different shapes, and to use the formula to find the volume.

The relationship between the circumference and diameter of circles

Case 5.1 (Before teachers try this with their pupils they will probably like to try the experiment in their INSET group.) Some 10-year-olds made a 'grid' graph of this relationship (Fig. 5.19). First, they made a large collection of circular objects, and marked out the diameter of each object in turn on the horizontal axis, starting at zero each time. Then they made a collar of coloured Sellotape to fit the circumference of each object exactly. They fixed this circumference 'vertically' at the other end of the corresponding diameter. When the graph was displayed on the wall, Bob said, 'I think I can see a relationship.' He took a piece of string and stretched it along the tops of the circumference columns. The pupils were excited to find that the tops of the columns were very nearly in a straight line, and even more so when they found that the string line went through zero.

After some discussion the teacher asked her pupils to plot the 3-times and the 4-times tables on the graph. (During the previous week they had plotted the multiplication tables from 1 to 10 on one grid.) Then she asked them whether the string graph of the circumference/diameter was nearer the 3-times or the 4-times multiplication table graph. They replied,

> The circumference/diameter graph is much nearer the 3-times table graph. So not much more than 3 diameters will make the circumference.

They began their calculations by taking 3 times the diameters to make the circumference; they also found the approximate diameters of circles when they only knew the circumference, by dividing the circumference by 3. Later on, they used more accurate approximations for π (3.142).

Area of a circle and volume of a cylinder

The formula for the area of a circle: $A = \pi r^2$, was developed in Chapter 3. The pupils should be given a variety of calculations using the formula. For instance, calculate the area of circles with radius (i) 25 cm, (ii) 6.7 m, (iii) 64.5 m. Use $\pi=3.142$.

A cylinder is a prism with circular cross-section. The formula for its volume V is therefore

V = area of base $\times$ height

or $V = \pi r^2 h$

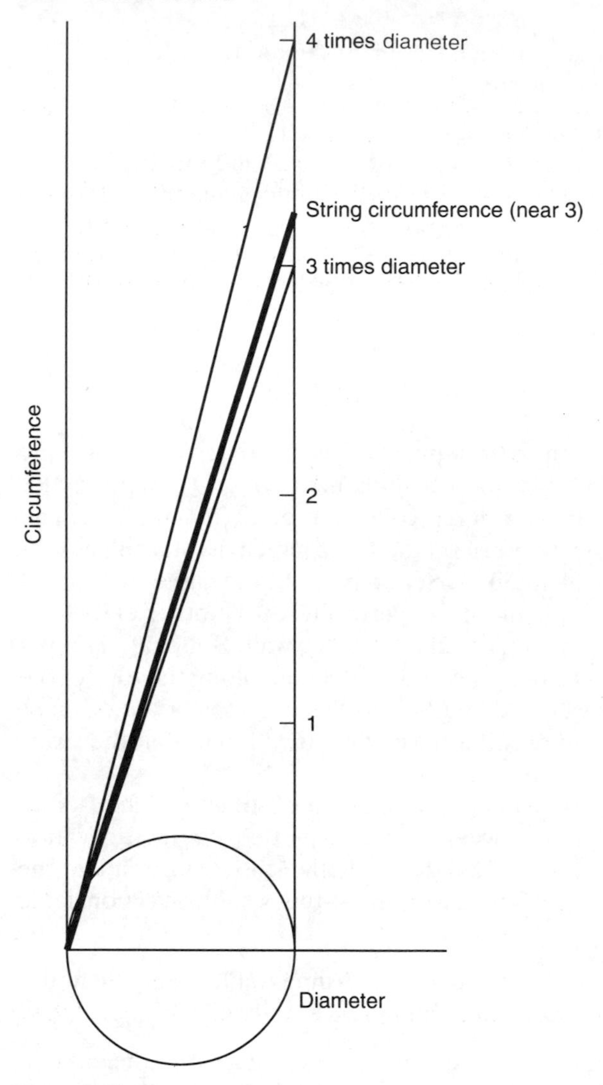

Figure 5.19 *The circumference/diameter ratio of a circle*

Give the pupils calculations, such as to calculate the external volume:

1. of a cylindrical tin 6 cm in diameter and 16 cm high,
2. of a drainpipe 1.3 dm in diameter and 2 m long,
3. of a cylindrical cup, 6.4 cm in diameter and 8.5 cm in height.

Ask the pupils to measure a variety of solid cylinders and to use the formula for finding their volumes. Sometimes suggest that they make up some interesting problems for the class book. Encourage them to use their calculators.

Areas of plane figures

The teachers (and pupils) have already found the formula for the area of a rectangle:

area = base × height

So it can be seen that in Figure 5.20(a) the area of rectangle ABCD = AB × BC, i.e. $6 \times 4\ \text{cm}^2 = 24\ \text{cm}^2$.

From this simple formula, we can then find the area of other plane shapes: triangles and parallelograms. We begin with a right-angled triangle ABC (in Fig. 5.20(a)).

$$\begin{aligned}\text{area of triangle ABC} &= \tfrac{1}{2} \times \text{area ABCD}\\ &= \tfrac{1}{2} \times 6 \times 4\ \text{cm}^2\\ &= 12\ \text{cm}^2\end{aligned}$$

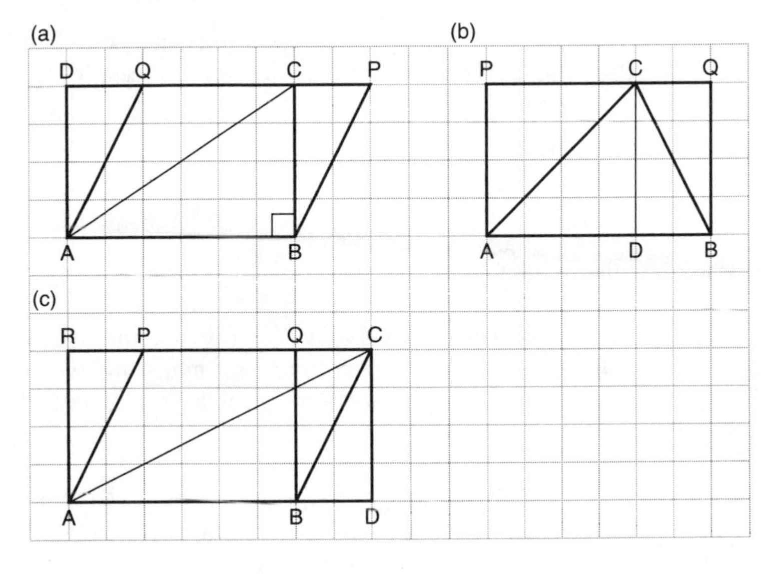

Figure 5.20

The triangle ABC in Figure 5.20(b) can be split into triangle ACD plus triangle DCB. So

$$\begin{aligned}\text{area of triangle ABC} &= \tfrac{1}{2}\ \text{ADCP} + \tfrac{1}{2}\ \text{DBQC}\\ &= \tfrac{1}{2}\ \text{ABQP}\\ &= \tfrac{1}{2} \times 6 \times 4\ \text{cm}^2\end{aligned}$$

In Figure 5.20(c), obtuse-angled triangle ABC has the same area as triangle ADC less that of triangle BDC, so

$$\begin{aligned}\text{area of triangle ABC} &= \tfrac{1}{2}\ \text{ADCR} - \tfrac{1}{2}\ \text{BDCQ}\\ &= \tfrac{1}{2}\ \text{ABQR}\\ &= \tfrac{1}{2} \times 6 \times 4\ \text{cm}^2\\ &= 12\ \text{cm}^2\end{aligned}$$

We notice that the areas of all three triangles conform to the formula

area = $\frac{1}{2}$ base × height

When introducing these activities to pupils we suggest that they can draw the shapes on squared (cm) paper, cut them out, and check the formulae by folding. For example, in

Figure 5.20(b), if triangle ACP is folded about diagonal AC, and triangle BQC is folded about diagonal CB, the two triangles will coincide with triangle ABC and will be seen to be half the rectangle ABQP. The obtuse-angled triangle is more difficult to solve pictorially – but let your able students work at it!

Area of parallelogram ABCP It is easy to show that the area of parallelogram ABCP is the same as that of the rectangle ABQR. Cutting off triangle APR they will find that side AP fits exactly on side BC. So when triangle APR is moved to position BCQ it completes rectangle ABQR. So the area of parallelogram ABCD is the same as the area of ABQR.

area of parallelogram = base × height

Provide a variety of numerical examples for the pupils, from which you can assess their knowledge and understanding. Also, encourage pupils to make up some examples for themselves.

KEY STAGE 3: SCALE

Enlarging a shape by a whole-number scale factor

Scale is an important topic with many applications in everyday life. The activities which follow will help you and your pupils to understand and to interpret maps of varying scales, but first make sure that you and your pupils are able to make sequences of squares, and can find the pattern of this sequence.

Activity 5.30 Provide identical plastic squares and ask groups of four pupils to make an ordered sequence of squares, using one colour for each square if possible. Then ask them to transform the sequence of squares into a block graph of the area of squares as in Figure 5.21. (This activity can be used as an assessment.)

The square sequence

Ask the pupils to describe the number pattern of square numbers. Some of them will notice the pattern from the 'rises' of the stairs in their area block graph. Some will volunteer that the differences are the odd numbers. Try to get a more precise answer – perhaps 'odd numbers beginning with 3'. If so, ask them where the first odd number (1) is. This may suggest that if we start with the number zero, the square of zero must be zero. (A 9-year-old once said, 'If a square hasn't got a side it hasn't got an area either!') The difference between 0 and 1 is 1, which gives us the first odd number.

The sequence of numbers is: 0 1 2 3 4 5 6 . . .
The sequence of squares is: 0 1 4 9 16 25 36 . . .
The differences are: 1 3 5 7 9 11 . . .

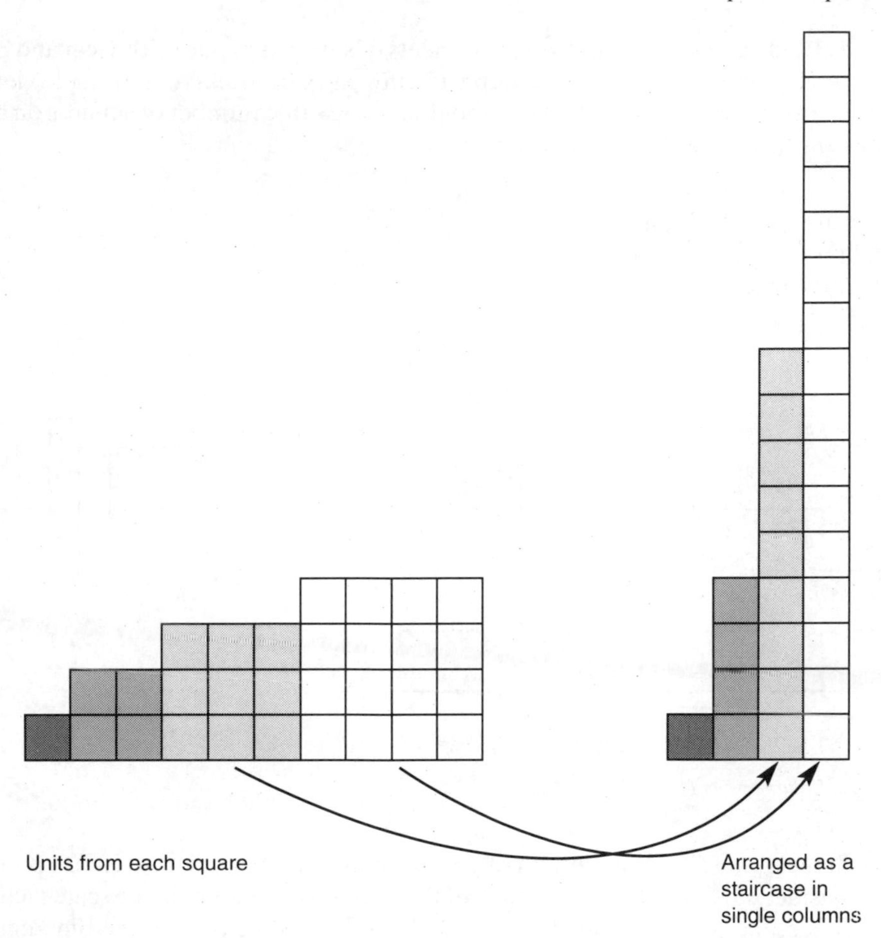

Figure 5.21 *A sequence of squares transformed into a block graph*

The cube sequence Some of the pupils may be ready to tackle three-dimensional scale. For this activity provide identical cubes and ask your pupils to make a sequence of cubes, and to look for a number pattern. (Let them use calculators.) If they come to a dead end, remind them that 'In mathematics we stop at nothing!' Then recommend that they continue finding subsequent differences.

The sequence of numbers is:	0		1		2		3		4		5		6 . . .
The sequence of cubes is:	0		1		8		27		64		125		216 . . .
The first difference is:		1		7		19		37		61		91 . . .	
The second difference is:			6		12		18		24		30 . . .		

Investigating scale

Children can begin to investigate scale at Key Stage 2.

Activity 5.31 Provide each child with two sheets of squared paper, with 1-cm and 2-cm squares. Ask them to draw the same simple picture on both sheets, e.g. the side view of a lorry or car, or a letter such as L. They should use the same number of squares on both their pictures (Fig. 5.22). Ask:

- How are your pictures the same? (shape)
- How are they different?
- What is the scale factor?
- What is the ratio of the perimeters?
- What is the ratio of the areas? (1 to 4)

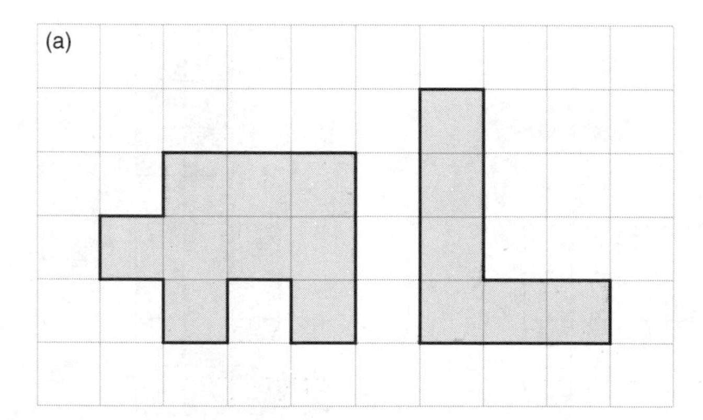

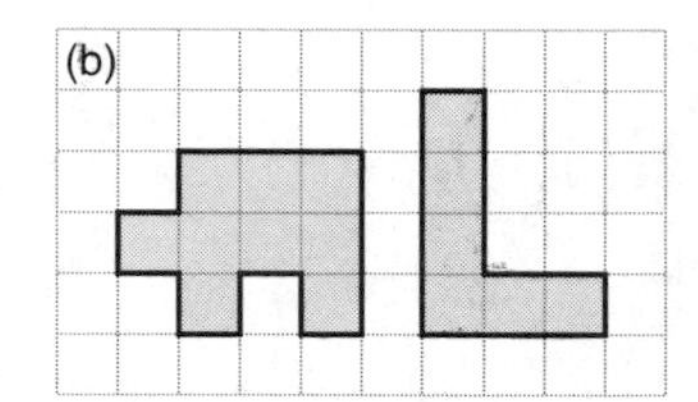

Figure 5.22 *Enlargements (Level 3)*

At Key Stage 3, the simplest method of enlargement is to use any convenient point as 'centre', inside, outside, or on the outline of the shape. In Figure 5.23 the scale factor in each case is 2; the original shaded figure is enlarged to create the bold outline figure.

Activity 5.32 Draw a polygon, take a centre and join it to the vertices. Enlarge the polygon by scale factor 2. Join the new vertices. Now try enlarging polygons with different numbers of sides. Take centres in different positions and use scale factors of 2 and 3. Discuss your methods. (Check the ratios of perimeters and areas each time.)

Activity 5.33 Ask pupils how they know that their scale factor is correct. Then ask them to investigate the scale factors of toy cars or model aircraft. Let them draw (using chalk) full-scale models of these toy cars or model aircraft in the playground and compare these with real ones.

Discuss the application of scale in geography, and what problems may arise when you are considering land area and road distances on the same map.

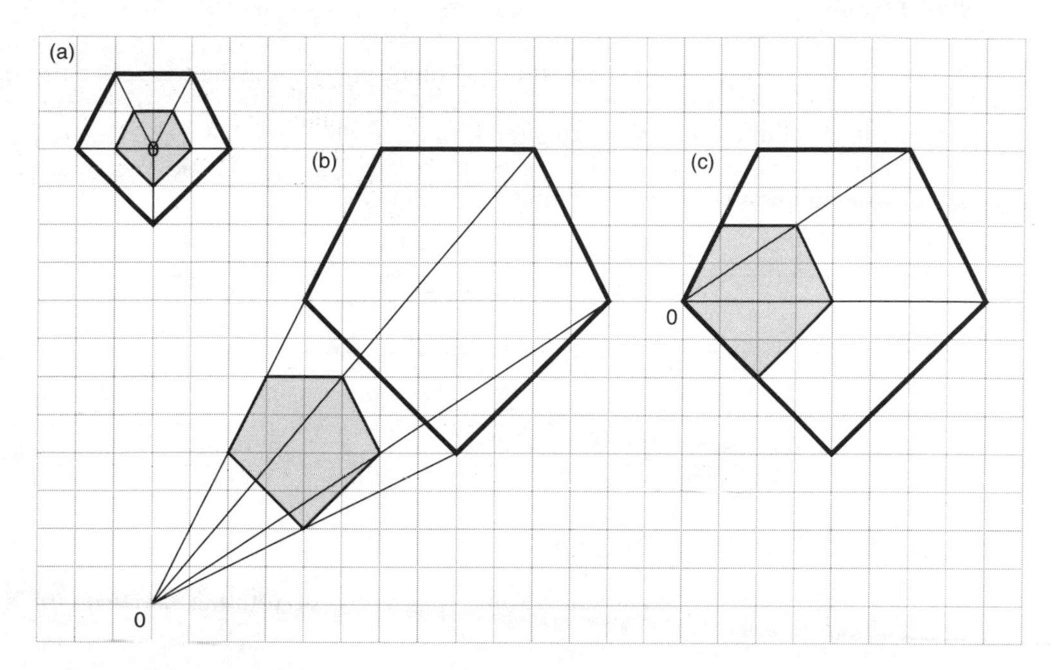

Figure 5.23 *Point enlargement (Level 4/5): (a) point inside; (b) point outside; (c) point at a vertex*

Classifying and defining types of quadrilaterals

There are a number of ways of classifying quadrilaterals.

Axes of symmetry

Sketch examples of quadrilaterals with at least one axis of symmetry. How many different types did you find? Figure 5.24 shows them all. How many axes of symmetry did each have? (Square: 4 axes; rectangle: 2; rhombus: 2; arrowhead: 1; kite: 1.)

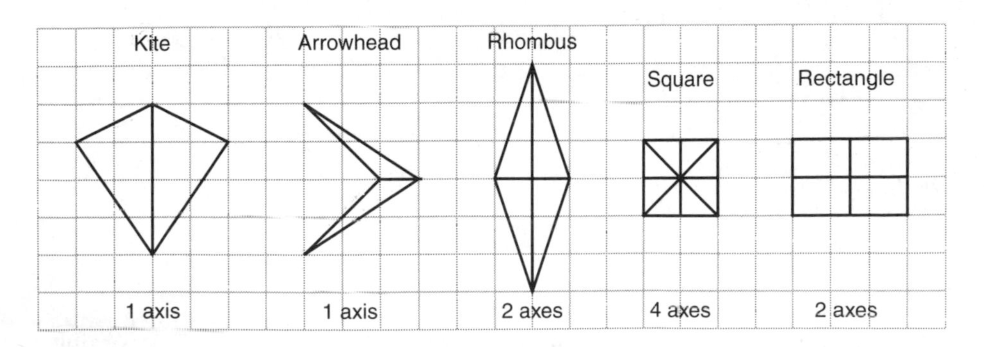

Figure 5.24 *Quadrilaterals with axes of symmetry*

Rotational symmetry

Sketch quadrilaterals with a centre of rotational symmetry, including the centre and the order (the number of different positions occupied by the quadrilateral in one revolution). You can check the order by putting a pin through the centre, and carrying out one revolution, after marking the starting position (see Fig. 5.25). Did you find 3 examples? (Square: 4; rectangle: 2; rhombus: 2; parallelogram: 2.)

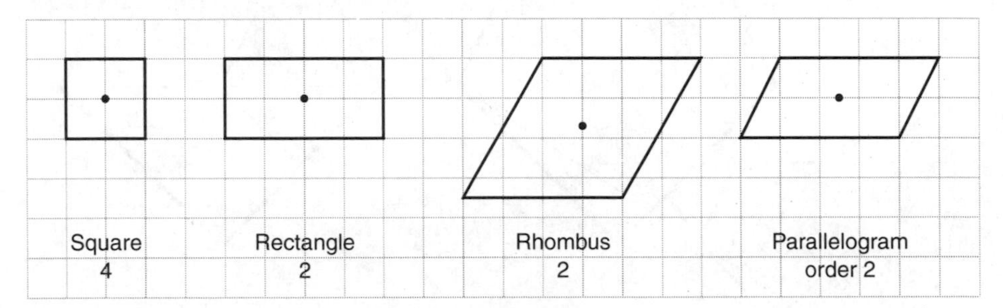

Figure 5.25 *Quadrilaterals with centres of rotation*

Activity 5.34 Three of the quadrilaterals have a centre of rotation *and* axes of symmetry. Which are they?

Tessellations

If several members of an INSET group draw congruent quadrilaterals they will be able to discover whether or not their quadrilaterals tessellate. Then construct congruent quadrilaterals with all their sides of different lengths. What is the sum of their angles? Do they tessellate? See Figure 5.26.

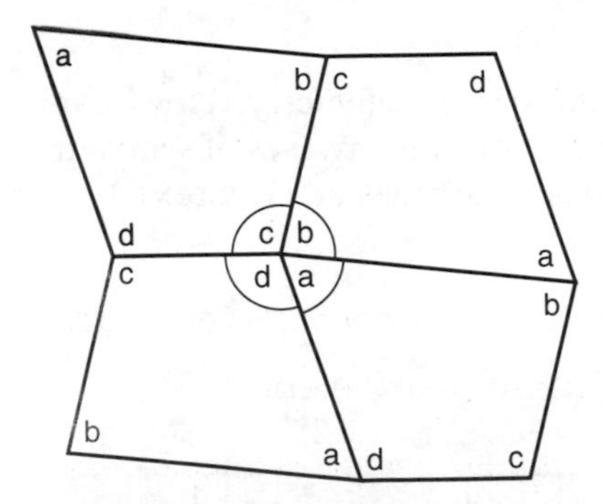

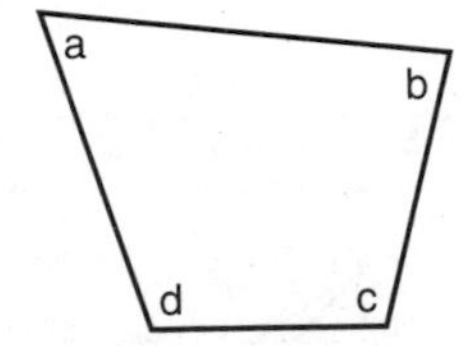

Figure 5.26 *Tessellation of quadrilaterals*

Definitions

Now look at the different definitions of the various types of quadrilateral. For example, a parallelogram is said to be a quadrilateral with:

- opposite sides parallel;
- pairs of opposite sides equal; and
- one pair of opposite sides equal and parallel.

Check that each of these separate definitions is sufficient. Then turning to a rhombus, is it sufficient to say: 'A rhombus is a parallelogram with adjacent sides equal?' What about a rectangle? Is it a parallelogram with one angle a right angle?

Although many of the practical activities in this section can be adapted for a number of the pupils in the final year of the primary school, fewer of them will be able to tackle definitions. They should, however, be able to list the properties of the different quadrilaterals.

Knowing and using angle and symmetry properties of quadrilaterals and other polygons

Supplementary angles

Your pupils have already found out that pairs of opposite angles of a parallelogram are equal. From this they may find out that angles *a* and *b* (in Fig. 5.27) are supplementary – either from equal corresponding angles or from remembering that angles *a* and *b* are supplementary because they are interior angles between parallel lines.

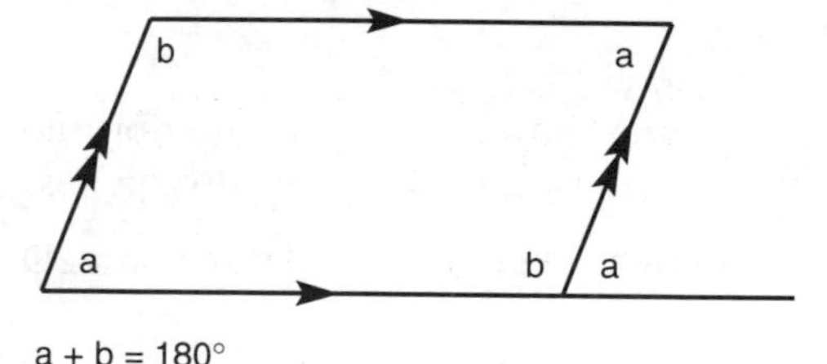

a + b = 180°

Figure 5.27 *Properties of a parallelogram*

Interior and exterior angles

Encourage your pupils to discover the sizes of the interior and exterior angles of regular polygons.

Activity 5.35 Figure 5.28 shows all the vertices of a regular pentagon joined to its centre.

- What are the angles at the centre?
- Ask your pupils to sketch a regular pentagon, to put in its centre and to join it to the vertices.
- Ask them to calculate the angles, giving reasons.
- Then ask them to record the interior and exterior angles of the pentagon.
- Ask them why these angles are supplementary.

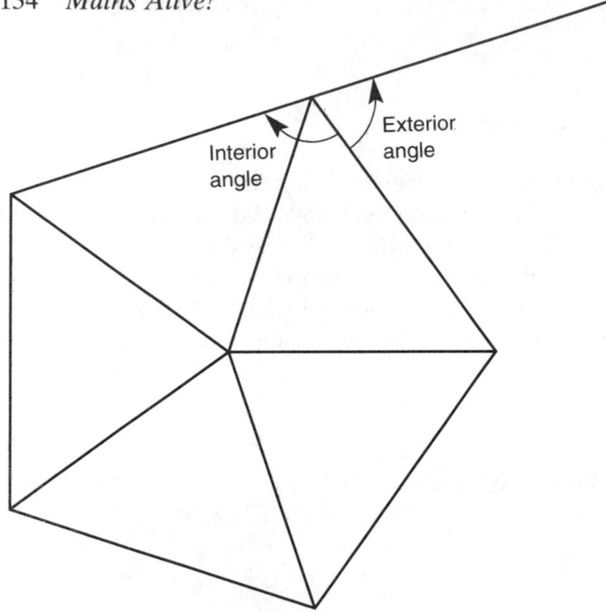

Figure 5.28 *Regular pentagon*

Repeat this activity for regular hexagons, octagons, nonagons and decagons.

- Ask them to make an ordered table showing the number of sides each regular polygon has, and its interior and exterior angles.
- Ask them to include the square and the equilateral triangle.
- Then ask them to plot the results from the table showing how the interior and exterior angles of regular polygons vary with the number of sides the polygon has.

Those pupils who have not yet used computer programs to generate and transform 2D shapes may like to do so at this stage.

Understand and use bearings to define directions (optional)

This topic has been moved to Geography. The activities which follow may well arise in preparation for a school journey – perhaps a geographical or environmental survey. The pupils should first learn how to use an orienteering compass, which measures angles clockwise from the North – the method used by ships and aircraft. Also, make sure that your pupils know the direction of North.

Direction finding

For practice in direction finding, use simple instructions to begin with, e.g.

- Walk 30 paces in a direction N 56° (Fig. 5.29(a)). Turn through 180°. What bearing will you use to return to the starting-point?
- Walk 50 metres on a bearing 130° from A to B, then 60 metres to C on a bearing 270° (Fig. 5.29(b)). What bearing will bring you straight back from C to A?

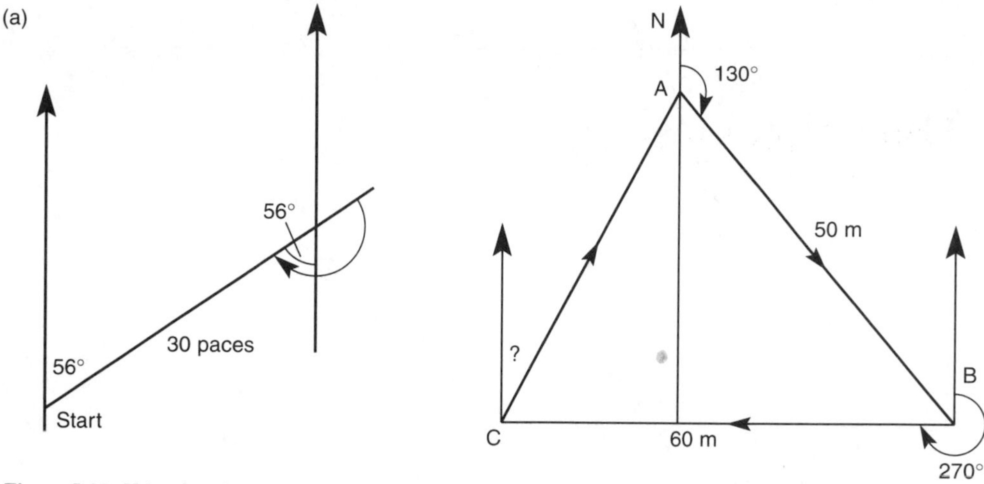

Figure 5.29 *Using bearings*

When your pupils understand direction finding, they can work in small groups on harder examples, which will also test their scale drawing abilities.

They can solve the following problems by making accurate scale drawings.

Activity 5.36 A pilot flies on a bearing of 12° for 80 km, then on bearing 312° for 50 km. Make an accurate drawing to find his distance from the starting point and the bearing he must take to get back to base.

Activity 5.37 A ship sails on a bearing of 205° for 10 miles, then on 220° for another 10 miles. Find the shortest distance back to port and the direction the ship should follow.

Activity 5.38 AB is a straight stretch of coast running due W–E, i.e. A is west of B. A lighthouse L has bearing 130° from A and 245° from B. The lighthouse is 4 km from A.

- How far is the lighthouse from the shore (shown dotted in Fig. 5.30)?
- A supply ship steams from A to L and on to B. How long is this journey altogether?

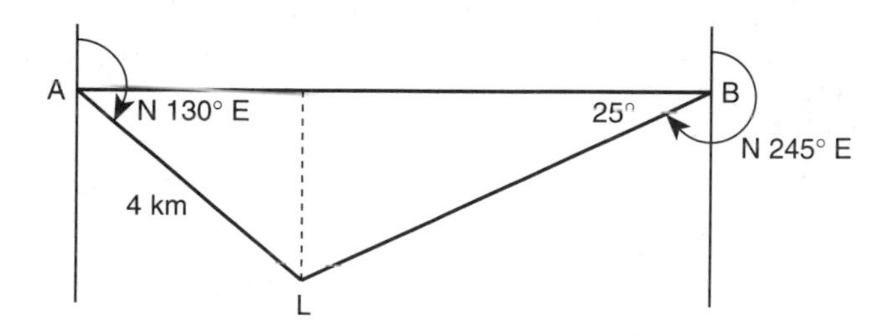

Figure 5.30 *Using bearings*

Chapter 6

Data Handling

INTRODUCTION

Pupils should collect, process and interpret data and should understand, estimate and use probabilities.

Until the publication of the National Curriculum, data handling and probability were not topics which primary teachers were accustomed to teach. Many pupils collected data, often on a limited range of topics such as favourite breakfast foods, colours or pets, but the data gathered were seldom fully exploited. In the past, many teachers did not have the opportunity of learning statistics, of analysing the data collected and presenting it in the variety of ways available today. This topic can be of great interest to pupils of different ages and should help them to understand some of the major problems today, e.g.:

- the need to conserve the world's resources of food and fuel;
- the need to provide the Third World with fresh water;
- the need to protect animals and plants on the endangered list.

It is important to realize that the work children will do in this section will be mostly of a practical nature. For this reason assessments are relatively easy for the teacher to make, by just observing the children as they apply themselves to their tasks and noticing particularly:

- how they carry out their tasks;
- how they explain what they are doing;
- how they respond to questions.

Probability has now been postponed to Key Stage 2.

A TRAFFIC SURVEY

What follows is an account of an actual project undertaken some years ago at a village school in which over 100 children took part. The children and their teachers planned to survey all vehicles which passed the school gate between 9.15 a.m. and 9.45 a.m. They chose their own materials for collecting and recording the data.

Problems to be addressed

Several problems could form the basis for a traffic survey.

- Which type of vehicle passes the school most often? (car, van, lorry, bus, bicycle)
- Does the frequency of each type of vehicle vary according to the time of day? (morning, afternoon, or 9.00–9.30 a.m., 11.00–11.30 a.m., etc.)
- Which make of car is most popular in this area/town/village?
- Do most cars carry passengers?
- Are white or coloured cars most popular in this area?

Data collection, recording and processing

We now consider the different methods of collecting, recording and processing data which were used at each level.

Case Study 6.1

Key Stage 1 The youngest children chose model cars, vans, buses and bicycles (all of different sizes) to represent the different vehicles. They arranged the models in sets and described their results:

> Two buses went past and two bicycles. There were many more cars.

The older infants used interlocking cubes of different colours to represent each type of vehicle. They made these into towers, each of one colour. They wrote about their findings:

> There were only two buses and two bicycles. Most cars went by, then vans, then lorries.

Key Stage 2 (Level 3) The youngest juniors used identical plastic squares of a different colour for each type of vehicle. They arranged these on a large sheet of paper labelled 'Our traffic survey'. They made a different column of squares for each type of vehicle, which they then labelled: cars, vans, lorries, buses, bicycles. Each column of squares started from the lower edge of the paper. (These children realized that to make comparisons they needed a starting line.) They placed the squares carefully, one above the other, without leaving gaps. When they were asked which vehicle had passed the school most often during the survey, they looked at their block graph to find the longest column. They then decided to do a second survey from 3.30 p.m. to 4 p.m. to see if the results were the same. They found that there were twice as many cars in their later survey.

> Perhaps more people work in the afternoon,
> they suggested.

Level 4 The oldest children used squared paper and constructed a block frequency graph. On the horizontal axis they labelled the vertical columns: cars, vans, lorries, buses, bicycles. Each time a vehicle passed the school they coloured one square in the appropriate column.

When they discussed their findings, Tim asked if the number of vehicles passing the school at noon, when the people employed at the pottery would be going home for lunch, would be the same as between 9.15 a.m. and 9.45 a.m. These children then made observations for another 30 minutes at noon, and found that many more vehicles passed the school at that time. In the subsequent discussion, they remembered that some people went to work at 8.00 a.m. – earlier than they had started their survey – so they had missed some vehicles in their first observations.

In the second survey they numbered the vertical axis to help them to read off the number of vehicles in each column. They had constructed a frequency diagram.

Assessment

The children's handling of this traffic survey shows how the same topic can be tackled at different levels. The teachers made sure that the different age groups had a clear understanding of the problems devised for their level. However, there is another important lesson to be learnt from this activity. For their first representations children use readily available three-dimensional material which may well be of different sizes. They are not troubled by this, nor are they concerned about a starting line. When they realize that for proper comparisons they need equal units and a base line, they are ready to undertake representations such as block graphs on squared paper.

These Level 1 assessments will help to establish whether the children are ready for this next stage. Any activity, for which the children have obtained definite results, will serve for assessment purposes. We shall use their favourite colour of T-shirt as an example. The child being tested had to collect from each of eight or more children a plastic square representing their favourite colour. He was then asked to make a graph with them on plain paper.

- Did he have a starting line?
- Were the units of one colour placed in the same column without leaving gaps between them?
- Did he identify as the favourite colour, the longest column?

Provide further experience as necessary, before introducing frequency diagrams.

A SCHOOL SURVEY

The next account considers general surveying tactics and statistical methods, and would be ideal for INSET sessions, to give teachers the opportunity to collect and represent personal data in a variety of ways. In the process, they may encounter ways of presenting data which will be new to them.

Questionnaires

The design of a questionnaire will depend on the problem being tackled. Initially pupils will generate their own problems rather than tackling a problem set for them, e.g. 'What is your favourite food?' Data collected from closed questions are easier to process, but it does mean that possible answers must be anticipated. The following questions are all examples of closed questions.

1. Are you male or female?
2. How many siblings were born in your parents' family: boys . . . girls . . .?
3. What is your birthday month . . . date . . .?
4. What is your telephone number?
5. Collect ten random car numbers (e.g. for LGN 188K the number is 188).
6. Do you live in a flat, a bungalow or a house?
7. What is your means of transport to school? By bus, by train, by car or walking?
8. What is your shoe size (UK sizes 1–12 – no half sizes)?
9. Choose your favourite fruit from: apple, peach, pear or strawberry.
10. Which sport do you prefer to play: football, tennis, rounders or cricket?

Personal measures

Personal measurements provide a wealth of data to study and give good practice in measuring and dealing with a variety of units. It is best to use metric tape measures.

Activity 6.1 Work with a partner and help each other to cut a set of strips, each in sugar paper and about 2 cm wide to match various measurements:

- height (shoes off),
- reach (arms outstretched sideways),
- perimeters of head, face, neck, waist,
- foot length,

Write your name and the name of the part measured at one end of each strip. Repeat this for the other person, using a different colour of sugar paper (to avoid confusion). First estimate, and then measure the length of these strips to the nearest centimetre and record the lengths in a table.

INSET

The data collected via questionnaires and/or measuring can be processed in a variety of ways and at different levels. Before the teachers attempt this in the classroom, it is important that they experience, at first hand, the problems that may arise.

Ask the teachers to work in groups of three or four, if possible at Key Stage 1 and then at Key Stages 2 and 3. They can then choose a topic, or two related topics, and collect the relevant information from all the questionnaires. Give them time to discuss what problems they would like to study, and to put the information in order. Here are some examples.

Birthdays

- Which is the most frequent birthday month?
- Is this the same for males and females?
- How many people are born on the first or last days of the months?

At this stage, nothing has been said about the size of the sample, but numbers can always be augmented by asking other members of teachers' households, or the pupils, until it is felt that 'enough' data have been collected, from which sensible conclusions can be drawn.

The measures

Many problems arise from these, e.g.

> Do tall people have a long reach?
> A 7-year-old once asked, 'Are you a square?' She carefully measured my height and reach, found that my reach was longer than my height and declared, 'No. You're the first wide rectangle I've measured. Your reach is longer than your height!'

Digit games

Suggest that they add the digits (single figures) of collected car numbers until they reach a single digit, e.g. for LGN 188K

$$1 + 8 + 8 = 17$$
$$1 + 7 = 8$$

Which single digit occurs most often?

Ask the teachers to represent the results in order, in a table, by a graph and in words wherever possible, and then make time to discuss all these tables, graphs and written comments. For the next INSET session ask the teachers to bring copies of the National Curriculum for May 1994 and September 1995, and the results collected from the questionnaires, augmented if necessary.

Vocabulary As usual, new vocabulary is introduced at different levels, as well as concepts which may be new to the children and maybe to many teachers. We now consider this in the light of data which might have been collected by the children if tackling the topic of 'our shoes'.

Key Stage 1

Activities can be concentrated on sorting objects, and dividing groups into sets of like things.

- Ask those children wearing one particular type of shoe, e.g. trainers, to stand in a group, and ask the others to sit on the floor.
- Then ask if there are more children wearing trainers or not wearing them.
- Ask them to suggest how they could find out.

They 'notice' which group has more children, and some children will be able to count how many more. Extend this concept, by asking them how many children are wearing, e.g. white shoes, or black shoes, etc.

Suppose the children had been asked to 'sort' the flower collection into those that were or were not red, and then also into those that had 5 or not 5 petals. The children could learn to represent these results by a **Carroll diagram** which combines both sets of data (colour, number of petals). If you mark out the four sections on the table top, each flower can be placed in the appropriate section. Figure 6.1 shows an example of a Carroll diagram, made by a group of 6-year-olds.

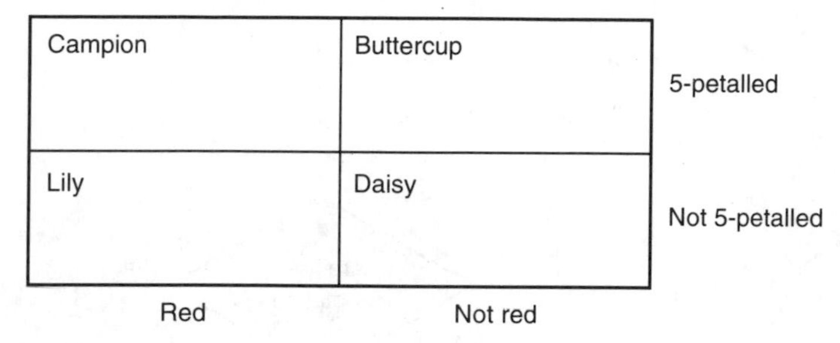

Figure 6.1 *Carroll diagram to sort flower collection*

Figure 6.2 is a Venn diagram showing the same data. Children often solve the problem of the overlap in Figure 6.2 (those red flowers which are not 5-petalled) for themselves if they have made Carroll diagrams first. Ask them to describe all the regions.

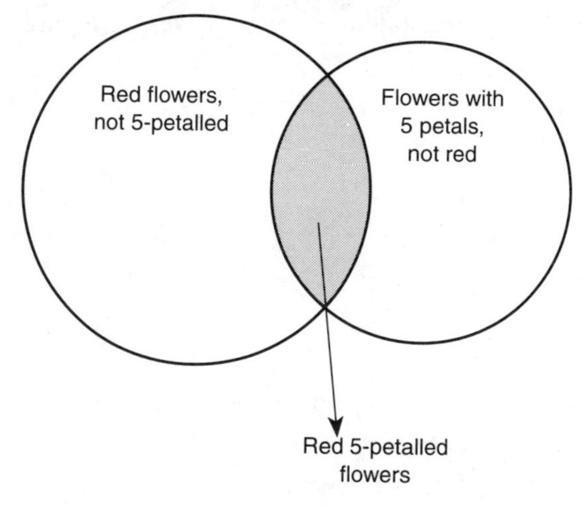

Figure 6.2 *Venn diagram (the two sets need not be the same size)*

Figure 6.3 is a **decision tree diagram** involving questions with yes or no answers.

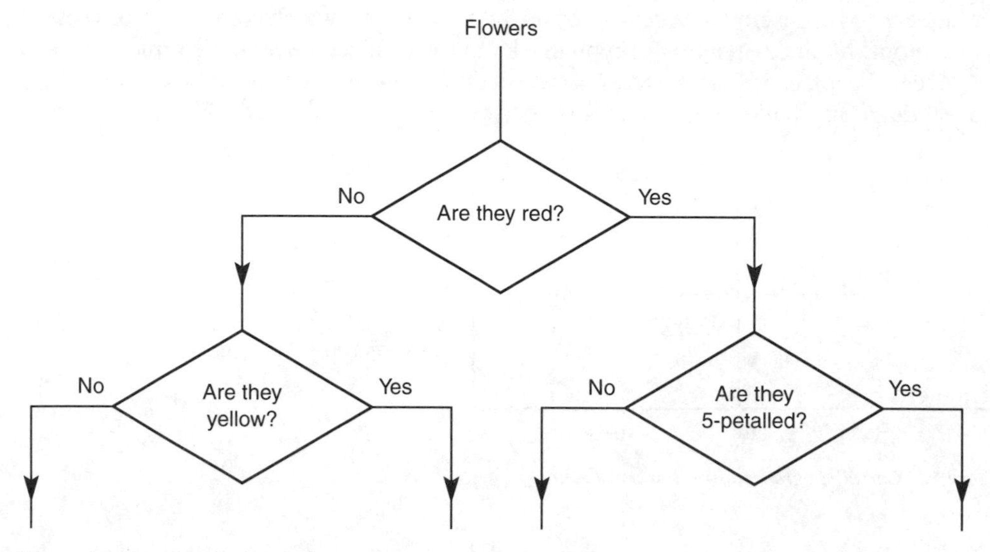

Figure 6.3 *Decision tree*

Relations and mappings (arrow diagrams) Arrows can be used for representation from Key Stage 1, when some children still find writing difficult.

A single arrow can be used to represent a specific statement such as 'wears'. Figure 6.4 shows:

- Peter wears trainers and football boots;
- Max wears all three types of shoe; and
- Jane wears trainers and sandals.

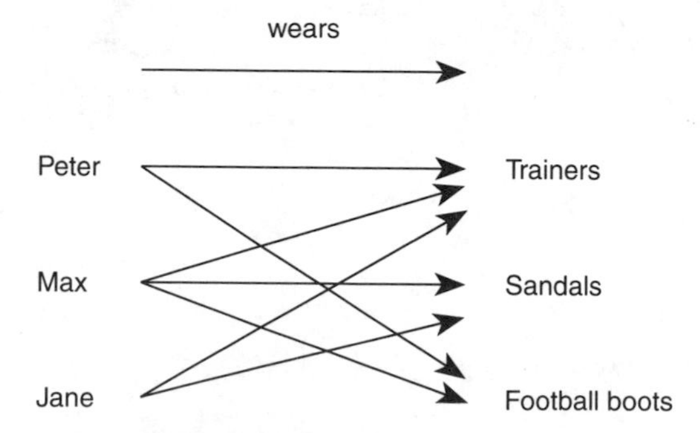

Figure 6.4 *Many-many relation*

Question the children:

- How many children wear football boots?
- Who does not wear them?
- Which type of shoe do all the children wear?

Some relations are **symmetric**, e.g. 'sits next to', because if A 'sits next to' B, then automatically B 'sits next to' A.

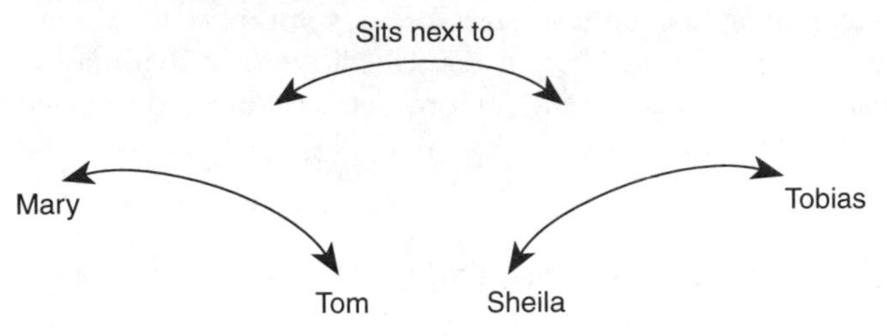

Figure 6.5 *Symmetric relation*

We could draw one arrow from Mary to Tom and then a second arrow from Tom to Mary. Instead, we draw one line with an arrowhead at each end. Can you understand why there is an arrowhead at each end?

Mappings

These are arrow diagrams in which the relationship is one to one. Many arithmetical operations have a one-to-one correspondence (see later for examples). For instance, if $\rightarrow$ stands for 'favourite fruit':

David's $\rightarrow$ is apple.
Ann's $\rightarrow$ is strawberry.
Jill's $\rightarrow$ is orange.

The children are different and each has a different favourite fruit.

Operations of arithmetic are also examples of one-to-one relationships or mappings, e.g. $\times$ 10, + 9.

$10 \times 1 \rightarrow 10$ $\quad 0 + 9 \rightarrow 9$
$10 \times 2 \rightarrow 20$ $\quad 1 + 9 \rightarrow 10$
$10 \times 3 \rightarrow 30$ $\quad 2 + 9 \rightarrow 11$

Discuss this in your INSET group. Both of these relations are called **mappings** (as in Fig. 6.6).

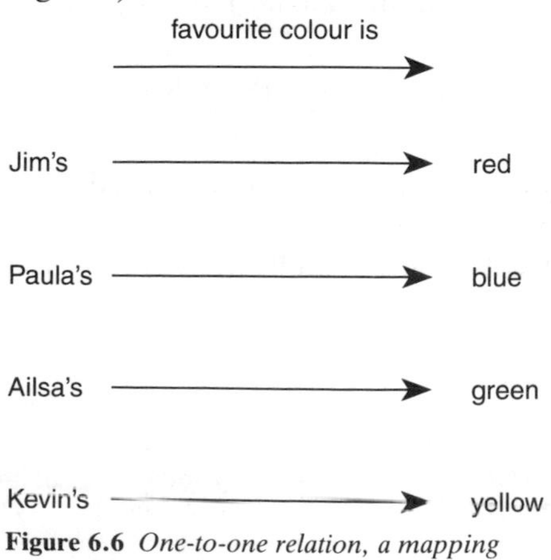

Figure 6.6 *One-to-one relation, a mapping*

Frequency graphs The children can be asked to find out their shoe size – as a whole number (half sizes should be avoided for now). The **range** of sizes included will depend on the age of your class. For 8-year-olds this might be sizes 2 to 6, so prepare labels and ask the children to arrange these labels in number order. Check with individuals to ensure that they all know the size of shoe they wear, before you ask them to put one shoe – their left one, say – into the correct size set.

- Which set contains most shoes?

The children may need to make a three-dimensional 'graph' with the shoes, to help them answer this question. Some may use one wall as a starting line for all the sizes. Others may realize that the shoes should be arranged one after the other without leaving gaps. When the 'floor' graph is finished, ask further questions.

- How many shoes are there altogether?
- Is there any set with zero shoes?
- Do any sets have the same number of shoes?

This may be the children's first experience of frequency graphs, so it is important that they are given time to absorb all the concepts involved.

Collecting data As children begin to collect data which involve larger numbers, they will find the 'gate' method of recording sets of five items more convenient. In this method, 1 stroke is recorded for every unit. IIII represents 4 units; a fifth diagonal stroke completes a set of five: ~~IIII~~.

At Key Stage 2 numbers can get larger still, and it is sometimes useful to invent a symbol to represent a group of, say, 10, 20, or even 50 units. If '+' represents 10, how many would '+ + + + + +' represent? If '–' represents 5 which number would be represented by '+ + + + –'? The children can devise examples for each other to represent.

Pupils at Key Stage 2 could be encouraged to collect shoe sizes (still in whole numbers) for boys and girls, eventually involving the whole year group. The data collected can then be used to make frequency graphs:

- for individual classes, separate graphs for boys and girls;
- for comparison purposes, a combined graph with separate data for boys and girls in the class (side by side);
- for the whole year group, frequency graphs for boys and girls separately and together.

In all these frequency graphs, shoe sizes are recorded on the horizontal axis and frequencies on the vertical axis. Ask the pupils first to discuss the graphs, then to write an account. Ask for a volunteer to describe the usefulness of the information collected to manufacturers of shoes.

Key Stage 2

> *Specify an issue for which data are needed. Collect, group, and order discrete data using tallying methods, with suitable equal class intervals, and create a frequency table for grouped data.*

Shoe sizes jump from one size (or half size) to the next and can only have specific values (lengths). There are gaps between the sizes – they are not continuous. Data of this kind are called **discrete**. Discrete data usually arise from counting. So when you measure the length of your foot with a metric ruler, you will find that the length only **approximates** to any one of the stated shoe sizes. Now, feet gradually grow through all the lengths from your shortest foot length to your longest. Lengths are therefore **continuous** data.

Case 6.2 A class of 10 year-olds decided to investigate the shoe sizes worn by those in their class. They wanted to find out whether the shoes the children were wearing were the right size for the length of their feet. First they collected the shoe sizes worn by every member of the class: a girl collected the shoe size worn by each girl and a boy collected the boys' shoe sizes. Then they made a table of these (Table 6.1) and a block graph (Fig. 6.7).

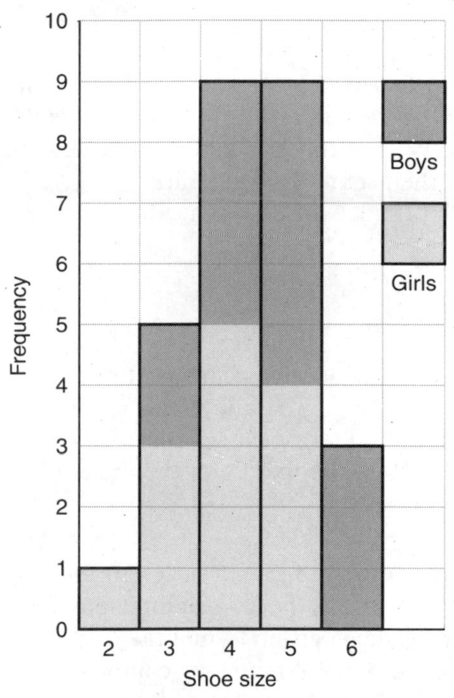

Figure 6.7 *Frequency graph of shoe sizes for 10-year-olds*

Table 6.1 *Shoe sizes worn by class 8*

Shoe sizes:	2	3	4	5	6
Number of girls	1	3	5	4	0
Number of boys	0	2	4	5	3

The pupils, in pairs, next measured their foot length. They discussed how accurate their measurement should be and decided to measure to the nearest centimetre 'because our feet grow so quickly'. Their data ranged from 17 cm to 27 cm, so they grouped their data:

15–19 cm 20–24 cm 25–29 cm

When they had made their table with grouped data, they realized that their groups were too wide and decided to regroup with 3 cm groups:

15–18 cm 19–22 cm 23–26 cm 27–30 cm

If some pupils are at Levels 3 or 4, they too might like to try this activity. (Some pupils also traced the outline of their shoes to check that their foot strip fitted within their shoe.)

KEY STAGE 2

Understand, calculate and use the mean and range of a set of data: understand and use the median and the mode in context.

Means, modes and medians What is generally understood by an 'average' (called the **mean**)? One infant teacher described an average as 'fair shares for all', and that is the simplest description of an average to be found. We use this 'mean' later to calculate average foot lengths: add up all the lengths and then share the total fairly.

Looking at the numbers in the frequency table of shoe sizes, though, which size best *represents* the set of numbers for the girls?

- Would it be size 4?
- Which best describes the boys' sizes?

If you are a 'buyer' for a shoe shop chain, the most frequently occurring statistic (the **mode**) is the most useful size to know. At other times, it may be interesting to note the 'middlemost' or **median** value. If there are an odd number of classes, this will be the middle one. When there is an even number of classes, the middle of the middle two classes is taken to be representative of the group.

INSET activity Working in groups of four, use your foot-length strips to find the average foot length of the group. Establish which group has the longest total length for foot strips (Sellotape should be available). Then ask each group to find the mean foot length of their group. Observe how they do this. Do they fold the long combined foot strip into four (quarters)? Ask them to describe exactly how they found the mean: first added the strips together, then shared the total fairly. Suggest that they try this activity with their own pupils, but before asking them to find the mean foot length ask:

Suppose all four of you had the same foot length, how long would this be?

More on means, modes and medians Pupils need to experience more than one activity on this topic, which is often learned without understanding. Therefore, provide a wide range of measuring activities within which the pupils can find the range (group and class) and the mean. Each time ask them the 'key' question to help them to understand what an average is. For example, if each group is finding the average time that members of this group take to run 80 metres, ask:

If you all ran the race in the same time, how long would you take?

Discuss the methods they choose to use.

Activity 6.2 First, with the pupils working in groups, ask them to measure their height strips. Ask them what would be a sensible degree of accuracy when they are measuring in centimetres. Correct to one centimetre?

Ask each group to find out the range of height lengths within their group. Suggest that they arrange the heights of all the class in height order, and state the range of heights for the whole class.

Then they can find the median (middle value) for the height, and the mode (the height which occurs most often).

Next, they can decide on their own group's average and whose height is nearest to this group average. Ask them to find the mean height for the whole class by including all of their height measures. Which pupil's height is nearest to the class's mean height?

We shall now consider the remaining Programmes of Study for Key Stages 2 and 3. It is hoped that nearly all pupils will cover Programmes of Study from Levels 1 to 4. Not many pupils are expected to complete Key Stage 3, but some will enjoy starting on the work for this Key Stage, and a few will complete this stage.

Key Stage 2

Entering and accessing information in a simple data base: e.g. a card data base.

Some teachers may not have used a computer database, so making a card database should help them to understand how the computerized database works. Suppose the card database holds information of all the staff in the school (collected from the questionnaires created earlier), then there will be one card for each teacher, on which is written their name and any other relevant information. Some data can be represented by e.g. 'is female', 'lives in a flat', 'has a brother', and so on. For this type of data, columns are made on every card to represent each category and a hole is punched at the top of each column (Fig. 6.8). If a teacher does not possess that attribute, the card above the punched hole is cut away. If the teacher does possess the attribute, the card and the hole are left intact.

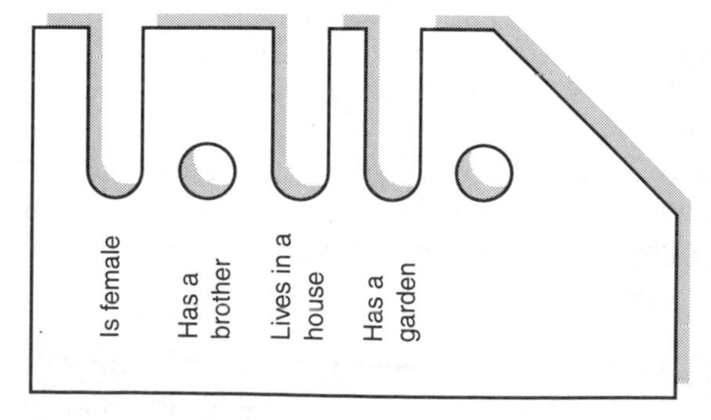

Figure 6.8 *Card system*

To extract the cards for the male teachers, a knitting needle is pushed through the holes in the first column. Cards for male teachers remain on the needle, while cards for females fall away. To extract cards of males who live in a flat, we need 'males' *and* 'live in a flat'. For this, the knitting needle is pushed through the second column using only those cards for males extracted earlier.

Constructing and interpreting bar charts, and graphs (pictograms) where the symbol represents a group of units.

The teachers should compile a list of enquiries, collected from the pupils, which they would be interested to investigate. Weather observations such as daily hours of sunshine, rainfall, wind direction and speed, are all good sources.

Pupils should extract specific pieces of information from tables and lists

Teachers and their pupils have already had experience of extracting and processing information from the questionnaires. In addition, pupils should study local train and bus timetables. Make sure that the pupils use the 24-hour clock, and ask them questions such as:

- Which is the fastest train of the day to your nearest large town?
- Which is the slowest?
- How long does each train take?
- Find the first and the last trains.
- How many trains are there on weekdays?
- How many are there on Sundays?

Sale catalogues are also useful sources of data, e.g. for the pupils to find where they could buy the cheapest Walkman.

Ratio Ratio can be introduced by using the measures recorded earlier to find approximate ratios.

Begin by comparing height and reach strips. The teachers can then assign themselves to one of three sets: squares, tall rectangles and wide rectangles. They should discuss their findings, and display them in an appropriate format. Then they can compare head and face perimeters, and finally, waist and neck perimeters. What do the teachers discover about this last ratio? Can they find another 'body' ratio which approximates to 2:1?

Making graphs First ask the teachers to make a graph using all personal strips in order of length (longest on the left). They can discuss how the order of their strips differs from those of their colleagues.

Next, each teacher is asked to choose a type of strip (e.g. perimeter of head) and to collect all these strips from their colleagues. They arrange their type of strips in order of length with the longest on the left. Arrange these graphs in a line along the floor. Then ask for comments and record these.

Scale models Some teachers may like to use their strips to make a full- or half-scale (3D) model of themselves by stapling their strips together, and stuffing the model with crumpled newspaper.

- How did they obtain half (or quarter) measures?
- What does halving do to the area?
- What does it do to the volume?

Key Stage 3

> *Constructing and interpreting frequency diagrams and choosing class intervals for a continuous variable.*

The teachers can use their height strips to help them to find suitable class intervals for their heights. Ask them to arrange their height strips in order of length. We shall take a range of 145 cm to 180 cm, and class intervals of 5 cm. We have to organize the frequency diagram (Table 6.2) to avoid overlap (each height must occur in one class only).

Ask the teachers to check that their own height comes in one class interval only. Ask them then to make a frequency table of their own heights and to follow this with a block graph. They can compare their block graph with Figure 6.9.

Throughout Key Stage 3 emphasis is placed on pupils specifying issues for themselves:

- designing observation sheets and questionnaires (to survey opinion);
- collecting data, collating and analysing results;
- creating frequency tables for grouped data;
- inserting and interrogating data on a computer;
- drawing conclusions.

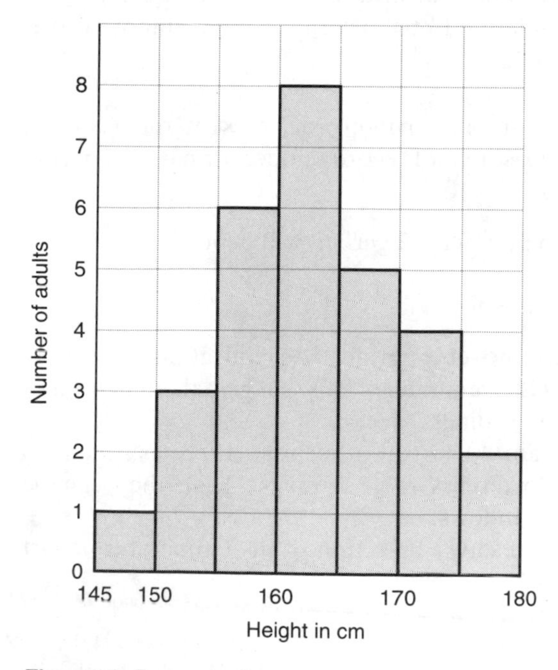

Figure 6.9 *Frequency diagram*

Table 6.2 Frequency diagram for heights of teachers

Class intervals in cm	*Tally*	*Frequency*
$145 \leq h < 150$	I	1
$150 \leq h < 155$	III	3
$155 \leq h < 160$	~~IIII~~ I	6
$160 \leq h < 165$	~~IIII~~ III	8
$165 \leq h < 170$	~~IIII~~	5
$170 \leq h < 175$	IIII	4
$175 \leq h < 180$	II	2

Encouragement is given to the inclusion of data arising on other subjects of the curriculum, e.g. investigating:

- weeds in a wild patch to find the most and least frequent plants;
- the favourite TV programmes in the class.

Construct and interpret a line graph and know that the intermediate values may or may not have a meaning.

It is always necessary, when drawing graphs, to consider whether lines joining points on a graph would have a meaning. Now, if a graph represents an algebraic relationship, for example, the squares of numbers, we can plot this as the relationship, $y = x^2$. The points on this graph ((1,1) (2,4) (3,9) (4,16) and so on) lie on a definite curve (called a **parabola**). All the points deriving from fractions, e.g. $\frac{1}{2} \times \frac{1}{2} = \frac{1}{4}$ will also lie on this curve (Figure 4.12).

However, if the points represent measurements of something that can vary randomly, as is often the case in temperature graphs during illness, intermediate 'readings' estimated from the graph might bear no relation to the patient's actual temperature at that time.

Activity 6.3 Before giving your pupils graphs to interpret, present them with a problem to solve in which they have to measure and record temperature over a period of time. For example:

> Where do you think the temperature in the school garden will be lowest?
> Where will it be the highest?
> Devise an experiment to solve this problem.

Suggest that temperatures need to be taken over a period. (As clinical thermometers are not easy to read, give them a Celsius thermometer.) This will give them first-hand experience of using graphs to record their findings.

Ask if any pupils have seen a temperature chart at a hospital. If so, ask them to describe it, and encourage the pupils to discuss their findings. Then they can be presented with a suitable problem such as follows.

Peter, aged 5, is in hospital. Figure 6.10 shows a section of his temperature chart from 22.00 hours, taken at intervals until 02.00 hours.

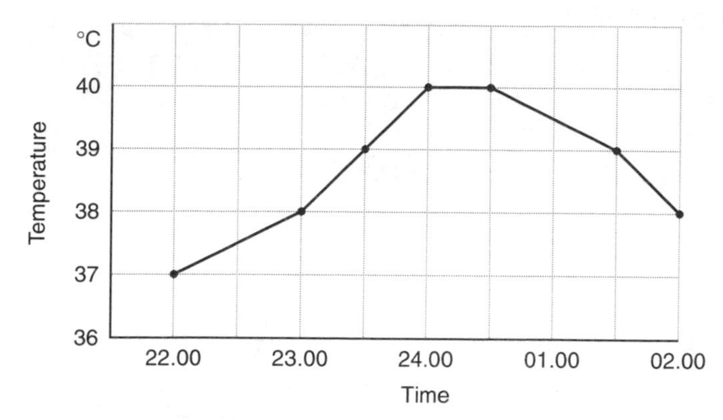

Figure 6.10 *Temperature chart for Peter, aged 5*

- What was Peter's highest temperature?
- At what time was Peter's temperature at its highest?
- At what time did his temperature start to go down?
- Describe what happened to Peter's temperature just after midnight, and just before midnight.
- Can you be sure that Peter's temperature was 37.5° at 22.30? Give the reasons for your answer.
- Why do you think that points marking temperatures are joined on charts?

Another type of graph in which intermediate points might or might not have a meaning is a **travel graph**. Time is normally recorded on the horizontal axis and distance travelled on the vertical axis. If the vehicle is travelling at a steady speed it will cover the same distance in the second minute as in the first. But this situation rarely happens; there are delays caused by traffic, crossroads, traffic lights, etc. and intermediate points are most unlikely to belong to the initial relationship. Points are therefore joined although we are aware that the speeds are average speeds and not exact speeds because of these delaying effects.

Activity 6.4 Figure 6.11 represents 11-year-old Roger's journey to return a video he borrowed from his cousin who lives 12 kilometres away. Roger leaves home on his bicycle at 5.00 p.m., cycles at 10 kmh for half an hour but then has a puncture which takes him half an hour to mend. He cycles the last 7 km at a speed of 12 kmh. He stays for 25 minutes, chatting with his cousin and then cycles home at 12 kmh. Make a graph to show Roger's journey and then check it against Figure 6.11.

- At what time did Roger arrive at his cousin's home?
- When did he get home?
- Find his average speed for the whole journey. (8 km per hour)

Pupils can invent their own journey stories and illustrate these by travel graphs. These can be used to exchange with other pupils.

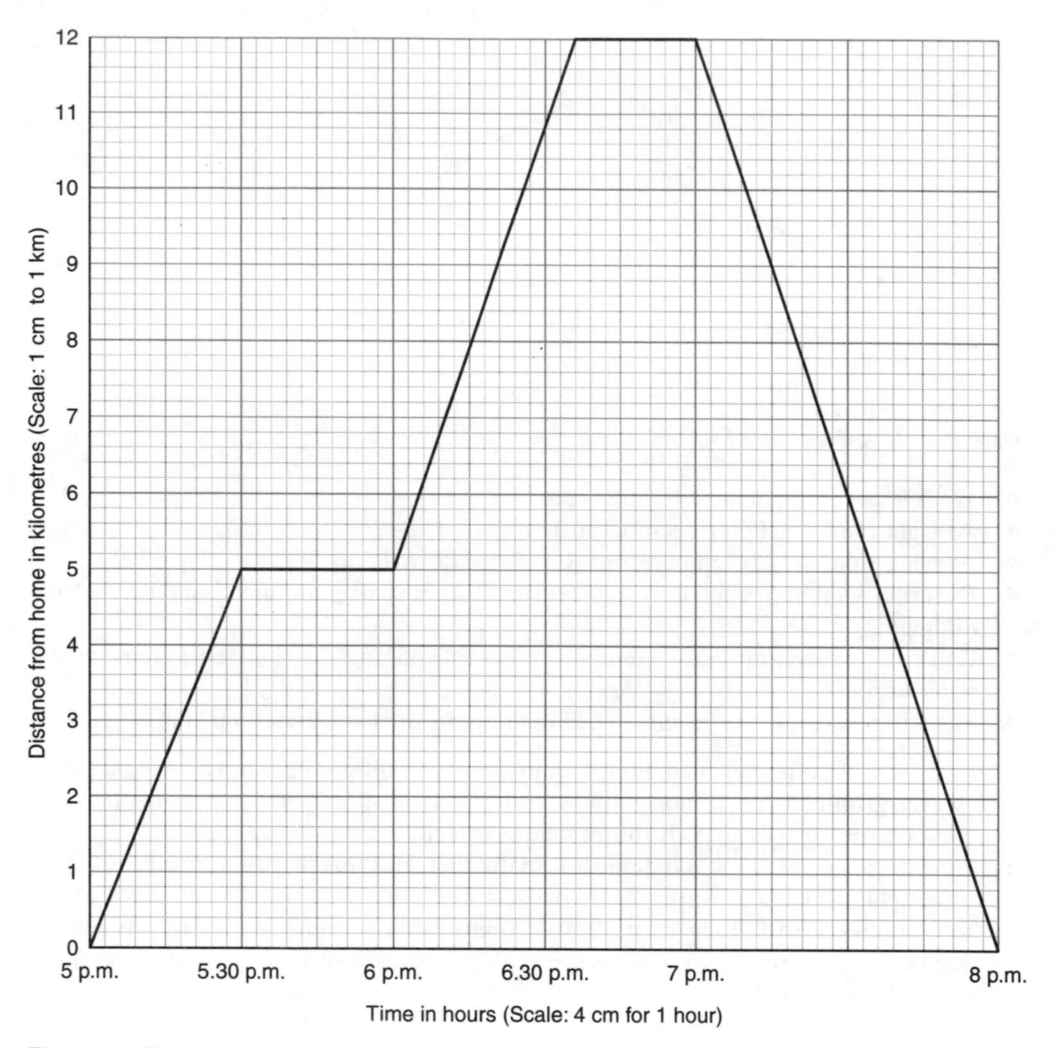

Figure 6.11 *Travel graph*

Construct and interpret further conversion graphs.

This topic is best introduced when the pupils are preparing to go on a school journey. They could collect the rates of exchange for £1 from the rates posted daily in high street banks. They should also find out the bank's commission charges.

Some pupils may be interested to study the fluctuations in the currency of the country they will be visiting.

A conversion table (as in Fig. 6.12) for converting Fahrenheit temperatures to Celsius can become useful when the pupils are asked to undertake a temperature study, using some Fahrenheit and some Celsius thermometers. The relationship is:

$$C = \tfrac{5}{9} \times (F - 32).$$

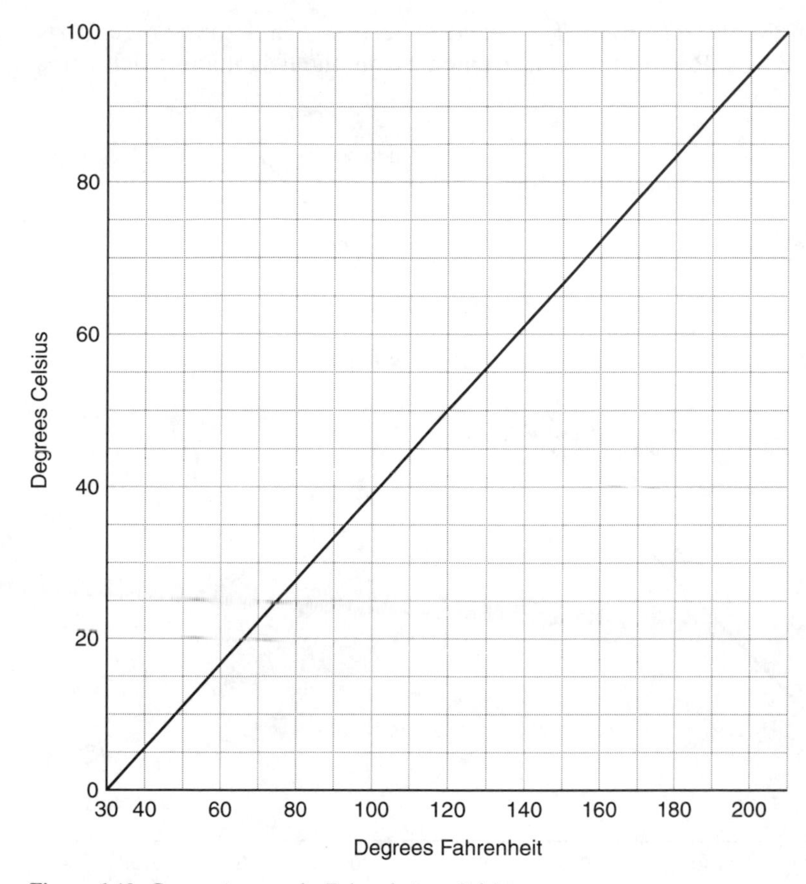

Figure 6.12 *Conversion graph: Fahrenheit to Celsius*

First find out how many approximate equivalents the pupils know, e.g. 61°F is approximately 16°C and 82°F is approximately 28°C. They may also know the freezing and boiling points of water in both scales (0°C, 32°F; 100°C, 212°F). Provide millimetre-squared paper and ask the pupils, in groups, to try to work out a conversion graph. 'Known' points can be plotted, and the children asked if they think that the graph will be a straight line. So should the points already plotted be joined?

Key Stage 3

Pupils should:

- *create scatter graphs for discrete and continuous variables and have a basic understanding of correlation;*
- *construct, describe and interpret information through two-way tables; and*
- *construct and interpret network diagrams which represent relationships or connections.*

Scatter graphs and correlation Correlation is a measure of the relationship between two variables, and scatter graphs offer one method of looking for such a link. For example, do you think that there is a close relationship (correlation) between height and reach? To investigate this, we can use the height and reach strips already prepared for every member of the INSET group. Two axes are needed: a 'horizontal reach' axis and a 'vertical height' axis. Record the measurements from the height and reach strips as shown in Figure 6.13.

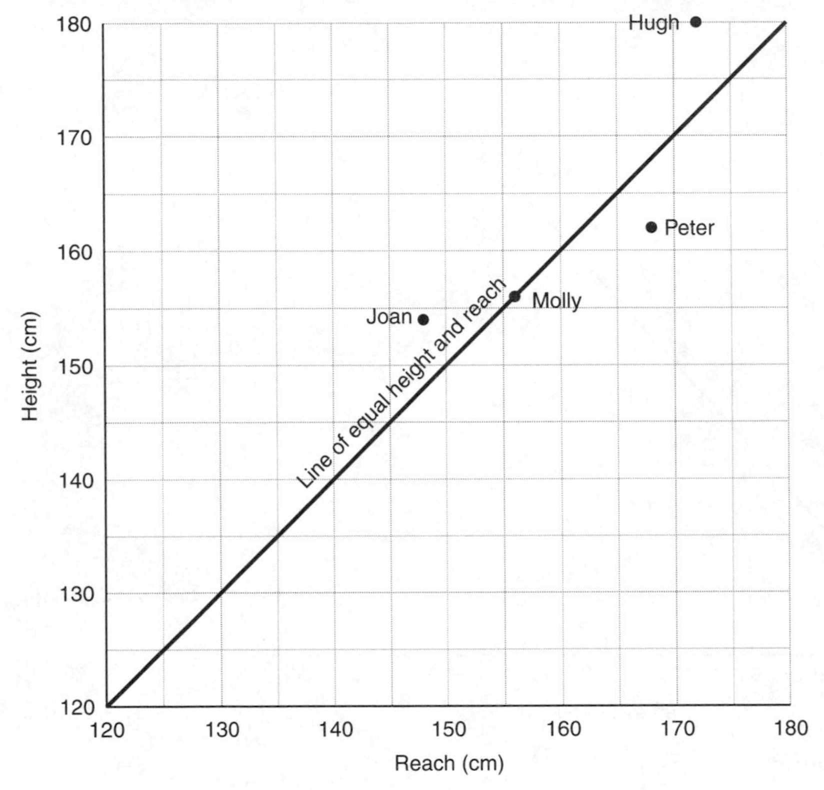

Figure 6.13 *Correlation between height and reach*

Do the points of intersection lie approximately on a straight line? Try to find a 'line of best fit' which seems to pass through the 'centre of gravity' of the points. (When this activity is done in the classroom, the pupils can also make a permanent scatter graph on graph paper after completing their floor scatter graph.)

- Do taller people have the longer reach in your scatter graph?
- What does your graph show about shorter people?
- What else does your scatter graph show?
- Is anyone in this group a square?

Two-way tables If any pupils seem ready to combine pairs of their personal statistics to study possible correlations between them, organize the previous activity with them. Other possible pairs of measures are:

- height and mass;
- face and head perimeters; and
- neck and waist perimeters.

Ask the pupils to suggest combinations which interest them. Ask them which combinations seem to show the closest correlation.

PROBABILITY

For the second part of this chapter, we consider activities on probability showing progression through Key Stages 1, 2 and 3. Probability is a new topic for most teachers (and pupils). Moreover, the activities provide different experiences of data handling and will therefore help to revise the concepts already encountered.

Until recently, few primary teachers in Britain included activities involving probability in their mathematics programme, but now that this topic is part of the National Curriculum, teachers are beginning to introduce it, even to 6-year olds, carefully recording the children's responses. Of course, activities involving probability are often familiar to young children because many of them will have played dice games, or card games such as *Snap* or *Beggar My Neighbour*, before they come to school. However, it is unlikely that parents will have discussed with them the probability of throwing a 6 at the next roll of the die (to start a game of *Ludo*, for example).

There are some decided advantages to the early introduction of probability activities in the mathematics programme. First, new vocabulary is being developed as children are asked to suggest some events which are certain to happen, some which are impossible, and others which might happen. The other advantages of the early introduction of probability are specific to mathematics.

- The activities are easy for children to understand. Because children enjoy playing games, their early experiences of mathematics can be very pleasant and varied. Moreover, teachers can use the games to help the children to memorize essential number facts.
- Probability experiments can provide opportunities for children to look for and discuss number patterns, and to make predictions.
- Probability activities also give children opportunities for organizing the data they collect in a variety of ways.

Because activities involving probability are all practical, assessments are not difficult for teachers to apply. In general, no special assessments need be made. By careful questioning (and we include some sample questions), teachers will learn whether the children understand what they are doing and have grasped what the activities were meant to teach them.

In the most recent publication of the National Curriculum, the introduction of probability has been postponed to Key Stage 2. However, this topic is very useful for enlarging children's vocabulary. Also, dice and card games help children to memorize number facts in a pleasant way. Teachers may therefore appreciate a few suggestions for the earlier stages.

Pupils should recognize possible outcomes of simple random events.

Young children ask many questions, some of which are easy to deal with because they have a precise answer. Others are more difficult because the outcome is indefinite:

- Will it rain this afternoon?
- Will it be fine on my birthday?
- Will the moon shine on Bonfire Night?
- Will the tide be out today?
- Will all my seeds come up?

Through the answers given to their questions, children gradually acquire a vocabulary to cover a variety of situations:

- perhaps
- likely
- unlikely
- certain to happen
- impossible

Key Stage 1

> *Pupils recognize that there is a degree of uncertainty about the outcome of some events and other events are certain or impossible.*

Activity 6.5 One way to start a group of young children on probability is by asking them to make sketches or lists of things they might take with them on a picnic. They could make three sets of things:

- those things they were certain to take;
- those they would certainly not take with them;
- those which they might or might not take.

(Six- and 7-year-olds draw pictures of the things they include in their three sets.) A group of 8-year-olds made lists as shown in Figure 6.14.

I am going on a picnic.

Things I would take

For certain	I might take	I won't take
rug	meat	piano
milk	apple	TV set
coke	ball	dishwasher
sandwiches	kite	
	book	
	radio	

Figure 6.14 *A picnic list*

The same 8-year-olds were then asked to decide, for each of the following statements, which word best described their likelihood.

- I shall be on time for the start of school tomorrow.
- My favourite pop-star will be visiting our school next week.
- I'll be a millionaire one day.
- Someone will cough in assembly tomorrow.

Then they were asked to make some statements for the others to judge, beginning with things which were certain to happen. Bianca said:

'I'm certain to be at school tomorrow.'

But her friends pointed out that she might be ill and then she would have to stay at home. Brian said:

'It's certain that school will be open tomorrow.'

All agreed. They found it easier to suggest things which could not happen. The first suggestion came from Sohail:

'It's impossible for my Dad to grow any more.'

Michael said:

'It's impossible for me to learn to drive tomorrow.'

Brian suggested:

'It would be impossible for me to go on a parachute tomorrow.'

Most of the suggestions for things which were possible but not certain were associated with the uncertain weather:

- It might rain tomorrow.
- We could have snow at Christmas.

After these preliminary sessions, some practical activities in probability were introduced with different emphases, in various age groups from 6 to 11.

Key Stage 2

Place events in order of 'likelihood' and use the appropriate words to identify the chance.

In the classroom the children can be asked to discuss the likelihood of a variety of statements, and to arrange the statements in order:

- very likely
- likely
- unlikely
- very unlikely.

It is important to vary the statements each time, and to encourage the children to make up some examples for themselves.

Understand and use the idea of 'evens' and say whether the events are more or less likely than this.

One of the early experiences children need to have is one in which the chances are even. The obvious choice is to present them with a coin and ask them: 'What is the chance of winning a toss?' However, when you try coin-tossing with children, you will find that many young children do not have the knack! It is useful to let them try this activity; some of them will be successful immediately and with practice the rest will soon follow.

As with most of the classroom activities, it is important for the teacher to have explored the investigation, within an INSET group, beforehand. So, within an INSET group, each of you tosses a single coin over and over again, recording the outcome each time, perhaps 50 times in all.

- Did you find that your scores of heads and tails were reasonably evenly distributed?
- Or were they very uneven?
- Did anyone score 49 tails and 1 head?
- If so, would you accept that you still have an even chance (1 out of 2) of tossing a head next time?
- Try another 50 tosses. Did you have better luck this time?
- Find the totals for your group and discuss the results.

This activity should serve to stress that although the symmetry of the problem (a coin has 2 sides so probability is $1/2$) seems to define the outcome, in practice the results may differ. The coin-tossing activity can be used with children, or if coin-tossing proves tricky, you can give them an alternative activity in which the chances are still even (1 out of 2).

Activity 6.6 For this activity you can use butter beans, marking or painting one side of each. The children toss 5 to 10 beans from a plastic cup and record each time how many painted and how many unpainted sides land uppermost. They can record their results in two columns labelled 'painted' and 'unpainted' on squared paper. Ask them to find and compare individual totals for painted and unpainted beans. Help them to find group and class totals.

(You can assess their ability with single-digit addition during this activity.) As an alternative, the children can toss small cubes with 3 faces painted red and 3 painted blue. Ask them what the chance is of throwing red.

Within an INSET group, the following investigation is useful. Each person tosses a die over and over again for 5 minutes, recording the score (1 to 6) each time. The teachers pool their results and find the totals for each score and see how close these are to an even distribution of the totals for each score, 1 to 6. They each then throw the die for another 5 minutes, recording the scores as before, and combining the scores from all members of the group once more.

- Are these results very different from the first set?
- If they amalgamate both sets of results, how close are the frequencies of occurrence of the scores 1 to 6?
- Are the totals for the scores 1 to 6 more evenly distributed than the original individual totals?
- Can you yet say what the chances are of throwing any particular number on a die?

Distinguish between fair and unfair

Give the children you teach a die to throw and ask whether they think they will throw a six next time, and why. Their replies can be very revealing. Young children often reply,

'Yes, because six is the biggest number on a die.'

Some say,

'No, I never get a six.'

At this stage, it is important that the children should come to realize that, if the die is fair (unbiased), each of the numbers one to six has the same chance of being thrown ($1/6$). To facilitate this realization, label a large sheet of paper with the six numbers on a die – as patterns of dots. Then each morning for a week, the first thing each child does is to toss a die 10 times, recording the score each time in the correct column. (Make sure that each child uses a plastic cup to hold the die when tossing, otherwise it is easy to obtain the same result over and over again!)

If the children are of reception age, they can use interlocking cubes and slot them together to compare the totals each day. Older children can toss the die many times and record on squared paper or by the gate method: IIII for 4 and ~~IIII~~ for 5.

By the end of the week, the totals should not be very different, and it is time to assess the children. Ask them if they are more likely to throw a 6 next time. Suppose they say,

'No, they all have the same chance.'

They have grasped the concept, and you can now ask them what that chance is. (You may need to repeat the activity if they do not answer your question correctly.) To check that they really do understand, ask them if they are more likely to throw a 5 (or any other number on the die, apart from 6). Ask them also how many times they think they will have to toss a die before they obtain a reasonably even distribution of all the scores 1, 2, up to 6?

This work can be extended further, but first the children must be familiar with odd and even numbers, and know how to distinguish between them without counting.

- If a single die is tossed over and over again, and each time you record whether the score is odd or even, do you expect more odd or more even numbers?

You can also give each pupil two identical dice:

- If two dice are tossed instead, and the sum of their scores recorded, what do you expect now?

Sampling events When statistics are being collected nationwide for a whole population, for example, the heights of 16-year-olds, it would be very expensive and time-consuming to measure the height of every 16-year-old pupil. Therefore, we take a structured sample (perhaps 5 per cent of the total, but this number varies). The sample must be random, taking into account such factors as geographical location and social class. By this means it was discovered that the mean height for 16-year-olds today is 2 inches more than it was 20 years ago.

Not all sampling experiments are nationwide. To introduce teachers to the prediction of the ratio of different coloured beads in a sampling bag, a small experiment was carried out.

Each set of three or four in an INSET group needs a thick sampling bag and identical beads of two colours. Each bag should have a drawstring. One person in each group decides on the number and colour of the beads to be put in the bag (perhaps equal numbers of both colours, the first time but then, different proportions). The beads are then well shaken, and another member of the group takes one bead from the bag and the colour is recorded. It is very important that the bead is returned to the bag, before the beads are shaken again, and another sample is taken. After 10 samples have been taken (and replaced), the members of each group consider the evidence before them and write down what they think the colours of the beads in their bag could be. Then the bag is opened to allow them to check. This activity can be varied sometimes using three or more colours, each time recording the results. The teachers should discover that, provided sufficient samples are taken, they can predict the contents of the bag.

Back in the classroom, it is interesting to observe children's responses to sampling experiments.

Case 6.3

Some 6-year-olds put beads of three different colours (20 blue, 8 red and 2 yellow) in a bag. With the teacher's help they made a written record.

> We wondered which colour would appear most often if we picked a bead at random, noted its colour and put it back in the bag.

Before they began sampling, the teacher asked the children to say which colour they thought would be picked most often. Here are their suggestions. These are quite revealing.

David	I think blue, because it is Leicester City's colour.
Paul	I think red, because I like that colour best.
Rachel	I think blue, because they are in a blue bag.
David	(having had time to think). We put the blue beads in first and they would go into the corners and when we shook the bag they would go into the middle and you would pick more out.
Celia	I think there would be more blue beads, because there are more in the bag.

The account continued:

> We tried to find out what would happen when we picked out a bead and replaced it 30 times.

The children included a tally of their results: 18 blue, 10 red, 2 yellow. Robert wrote:

> We were surprised that the results were nearly the same as the number of beads in the bag.

Young children respond in different ways to sampling experiments. Much depends on their previous experiences. Celia thought carefully before giving her answer, and David had time to think between his two answers. Because this topic is so new for young children we should accept all their answers, sometimes without comment, and certainly without criticism.

When you have tried sampling experiments with your children, you could compare their comments at your next INSET session. Equally, you could repeat this activity, using different materials, e.g. differently coloured cubes identical in size.

List all possible outcomes of an event.

Previous die-throwing activities should have established that there are six possible outcomes when tossing a die: 1, 2, 3, 4, 5 or 6. So, what is the chance of throwing a 6? Can the pupils say what the odds are *against* throwing any particular number on a die? (5 out of 6). If you wanted to work out the probability of throwing any number other than 6 when tossing dice, what would it be? ($5/6$).

Similarly, coin tossing should have resulted in 'heads' or 'tails', and so on. For each activity the children should be encouraged to list all possible outcomes, if only to help them record the results of the experiment!

Understand and use the probability scale 0 to 1.

Now we need to introduce the children to a 'chance' number line (Fig. 6.15), ranging from 0 (no chance) to 1 (certainty).

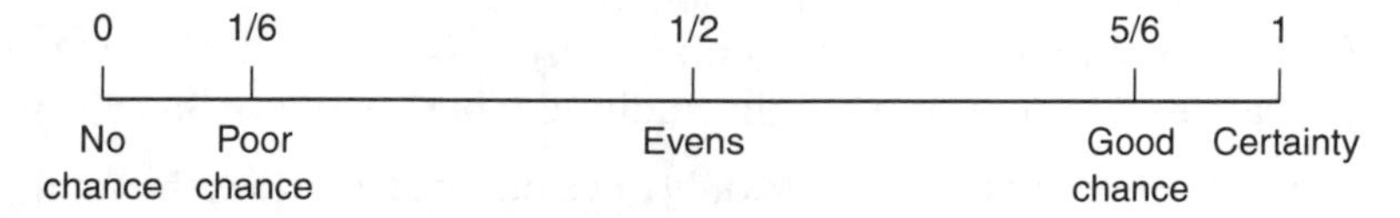

Figure 6.15 *Chance number line*

- Ask them to mark on the line where an even chance should come (halfway).
- Ask them to show you where the chance of tossing 6 on the die comes ($^1/_6$, poor).
- Where does the chance of tossing any number except 6 come? ($^5/_6$, good)

Giving and justifying subjective estimates of probability.

Ask the pupils to estimate, with reasons, the likelihood of rain tomorrow, and to express this as a percentage. Discuss the reasons given.

- We haven't had rain for a month so I think it will rain tomorrow.
- The forecasters said the probability of rain is 20% so I'll go for that.

There may be one or two pupils who will look at the weather map before making a decision. The weather does tend to be unpredictable, so a better example is to return to throwing dice. The next activity is always enjoyed by children, and makes a very suitable INSET activity.

Dice games

Activity 6.7 Every member of the group throws two dice at a time, adds the scores and records the totals. After 20 throws each, they identify the highest and the lowest totals achieved.

- What is the lowest number possible? ($1 + 1 = 2$)
- What is the highest number possible? ($6 + 6 = 12$)
- Is it possible to obtain every number between these two?

Each member of the group then prepares a recording sheet on squared paper with the possible numbers in order: from 2 to 12. They then throw their two dice repeatedly, recording the totals each time, in the appropriate column. They continue until one of the columns is complete. Individual results are discussed in the group, and then the results are combined.

- Which total turned up most often?
- Which total turned up least often?
- Have any of the totals the same chance of being thrown?

Now, for the next stage of the investigation, each member of the INSET group makes an addition square for the numbers 1 to 6 (Fig 4.1). Number patterns, including those on the diagonals, are discussed and described.

- Can you describe the pattern on the diagonal starting from the lower left-hand corner in more than one way? (We are adding pairs of equal numbers.) Table 1.
- How is this addition table like results of the experiment with two dice? (Both involve the addition of pairs of numbers from the set 1 to 6.)
- In the addition table, which total occurs most often?
- Which ones occur least often?
- Make a list, in number order, of the number of times each total occurs in the table.
- What is the frequency of the total 7 in the addition table?
- How many entries are there in the table altogether? (36)
- So, what is the probability of 7 appearing in the addition table? (6 out of 36 or $1/6$)

At the other end of the totals in the table, 2 occurs only once out of 36 (as does the total 12). We now aim to compare the frequencies of the totals for each number in the group's experimental results with those in the addition table. Using a calculator, we find $1/36$ is 0.028 (to 3 decimal places) and $1/6$ is 0.167. Work out the probabilities of the appearance of all the totals in the addition table, and arrange these probabilities in number order from the least likely to the most likely. Then we refer to the record of the total number of throws for the group and calculate

$$\text{Probability of throwing a total of 7} = \frac{(\text{no. of 7s obtained})}{(\text{total no. of throws})}$$

and use a calculator to convert this vulgar fraction to a decimal fraction. They can then compare the theoretical probability of throwing a total of 7 obtained from the addition table ($1/6$ or 0.167) with the results just obtained. This can be repeated for all the outcomes (2, 3, 4, . . . , 11, 12) to see how close the experimental results were to the theoretical results.

This dice activity is always enjoyed by children. Moreover it helps them to memorize important number facts. For example, the doubles of numbers 1 to 6; the near doubles, $1 + 2$, to $5 + 6$, and so on. Each time children are given this activity, it is important to tell them which set of facts is being tested. The addition table is better left until the pupils have further experience of dice addition. When they have made an addition table for pairs of numbers from the set 1 to 6, discuss the number patterns with them, before asking them to make a list, in number order, of the frequency of occurrence of all the totals in the table. Only when they are ready should children use calculators to find the probabilities of the occurrence in the addition table of all the totals: $1/36$, $1/6$, etc.

Activity 6.8 Vary the previous activity: this time throw two dice but record the difference between the scores. Work out the highest and the lowest differences achievable and prepare a record sheet on squared paper. Throw the dice for 5 or 10 minutes, recording the differences each time, and then combine the results from the group.

- Which difference occurred most often?
- Which occurred least often?
- Did any differences occur the same number of times?

Now make a difference table for the numbers 1 to 6; a difference table avoids introducing negative signs – the smaller number is always subtracted from the larger. Study the patterns in the difference table.

- Which difference occurs most often?
- Which difference occurs least often?
- What is the probability of 1 occurring in the table? (10/36)

Use a calculator to work out the probabilities (to 3 decimal places) for all the differences, and compare these probabilities with those from your experimental results. (You will need to divide the frequency of each difference in your experiment by the total number of tosses made by your group. If you express these as decimals to 3 decimal places, you can then compare these experimental results for each difference with those obtained from the difference table.)

To include negative numbers in the investigation, make a subtraction table instead, in which, for example, the numbers in the vertical column are always subtracted from those in the horizontal row.

Then use identical dice of two colours, red and blue say, and obtain the 'difference' each time by subtracting the score on the red die from that on the blue. Compare the two tables when completed.

In the classroom don't introduce the difference table until the children can make the addition table confidently; some children may get confused. Work through one or two differences to give them a start. When they have completed the difference table correctly, they can make a table (in number order) showing the frequency of occurrence of each difference in the table. They can then compare the probability of any one difference occurring in the table with the probability worked out from their experiment. Their successful completion of the difference table provides a reliable assessment of their ability to deal with differences.

This experiment should be reserved for older children who are familiar with positive and negative number lines. They could compare the probabilities obtained experimentally with those in the subtraction table. When trying these experiments with children, it is important to make time to discuss their responses with colleagues afterwards.

Card games

Activity 6.9 As a variation on dice games, you can use two sets of cards, numbered 1 to 6. Shuffle the double pack (of 12 cards), then take any two cards and add their scores. Record this sum, put the cards back in the pack and shuffle the pack again before taking the next pair. Continue with this activity (until you are bored?).

- Will the probabilities of the different totals occurring in the addition table be the same as in this activity?
- Compare your totals for the addition of the scores on two dice with those from the addition of the numbers on two cards.
- Similarly, compare the results for the *differences* between the scores on two dice and between the numbers on two cards.

Activity 6.10 For this game each player needs a set of cards, 0 to 12, arranged face up in number order in front of you. You play in pairs, and throw one pair of dice in turn. At each throw the player decides whether to add or subtract the scores, and then turn the card with that number face down. The first player to have all their cards face down is the winner.

- Can you work out the probabilities of throwing each of the numbers 0 to 12? (You will need to combine both addition and subtraction tables.)

When this activity is tried in the classroom, most children choose addition more often than subtraction, but they need to be encouraged to practise them equally.

Key Stage 3

Understanding that different outcomes may result from repeating an experiment.

The pupils have already discovered such variations when they were taking part in dice activities. Draw their attention to the variation in sampling activities.

Sampling and division The next few activities provide useful practice in sharing, the concept of division, odd and even numbers and remainders. As usual, it is best to try these activities within an INSET group before taking them to the classroom.

Activity 6.11 Place about 40 small pebbles (or shells or buttons) in a bowl. Take a sample (a good handful) and put it on a sheet of paper. Without actually counting, find out whether you have taken an odd or an even sample. (How did you do this?) Record whether your sample is odd or even, and then put the sample back in the bowl before taking another. Take ten samples in all.

- How many samples were odd?
- How many were even?
- Did you have the same number of odd and even samples, or more or fewer even samples?

Combine your results with others.

- How many samples were taken altogether?
- What is the chance of taking an even sample?
- What is the chance of taking an odd sample?
- If you, as a group, had taken the same number of odd and of even samples, what would that total have been? (How did you find out?)
- Which is nearer to this total, the number of odd samples, or the number of even samples?

Now, these two numbers should be the same, but very few adults (let alone children) think this can be true. So ask the group to take another ten samples and repeat their calculation. Repeat the basic questions:

- What is the probability of taking an even sample?
- What is the probability of taking an odd sample?

Back in the classroom, before introducing this activity you will need to give them practice in finding whether a sample is odd or even. Ask questions such as:

- Is there an even number of girls at school today?
- Of boys?
- Of boys and girls together?

Make sure that they take samples of a reasonable size. Give them plenty of practice before taking the next step. Make sure that they all realize that there is an even chance of taking an even or an odd sample.

Activity 6.12 This time the investigation centres on whether the pebble sample is divisible by 3 or not. So repeat the previous activity but record 'divisible by 3' or 'not divisible by 3' for each sample taken.

- What were your totals this time?
- What were your group totals?
- Were they nearly equal or decidedly different?
- Why do you think these numbers are so different? (Probably, the ratio is about $1/2$.)

Calculate an approximate ratio using

$$\frac{\text{the number of multiples of 3}}{\text{the number of non-multiples of 3}}$$

Now, take some more samples, divide each by 3 but this time, notice the remainders.

- How many possibilities are there when you divide by 3?

There are three altogether: 0, 1, and 2. Now take another 20 samples, but this time record your samples according to their remainders: 0, 1, 2. Then obtain the group totals as before, but for each remainder.

- Were the totals closer this time?
- What is the probability of taking a sample which is a multiple of 3?
- What is the probability of taking a non-multiple of 3?
- How many samples were taken altogether?
- Suppose that instead of your actual group totals for each remainder you had an equal number of samples for each remainder (0, 1, 2,), what would that number have been?
- Which actual total is nearest to that number?
- Do you know what that number is called? (It is an **average**, or an equal share.)

In the classroom, when this activity is used with the children, it is important not to omit any of the steps included here. The children need plenty of experience so that they know what to do when asked to take samples. They need practice in finding and recording the remainders when the samples are divided by different numbers. Make sure that they can explain the probability of taking a multiple (i.e. sharing with no remainder).

Activity 6.13 This activity can be used as an assessment, exactly as it stands. Ask the children to take 20 samples in all, to divide each sample by 4, and to record each result according to the remainder. Ask them to combine all their results. Ask them how close the results are. Then ask:

> If all four totals for each remainder had been the same, what would that total have been?

Observe how each child works this out.

- Does he find the total of the four numbers and then divide by 4?
- Ask him what this number is called (the average).
- Can he explain what an average is? (A fair share).

You can also use this activity as an example of both division situations. For instance, ask the children to put out two samples of 6 pebbles. Each set is to be divided by 3 in a different way:

- sharing among 3, we have 3 sets of 2;
- putting 6 into sets of 3, we have 2 sets of 3.

Activity 6.14 Each member of the INSET group needs sampling materials (pebbles/buttons etc.), and counters of four or more chosen colours (red, blue, green, yellow, etc.) Take a sample.

- Find out whether it is divisible by 2. If so, take a red counter and score 2 points.
- Find out whether the *same* sample is divisible by 3. If so, take a blue counter and score 3 points.
- If the same sample is a multiple of 4, take a green counter and score 4 points.
- For multiples of 5 take a yellow counter and score 5 points.

The game progresses with the players taking samples in turn and the player with the highest score wins.

- Did anyone have a sample divisible by 2, 3, 4, 5, and 6?
- What is the probability of taking such a sample?

Now, probability of taking a multiple of 2 is $1/2$, of taking a multiple of 3 is $1/3$, and of taking a multiple of 6 is $1/6$. We notice that the probability of taking a sample divisible by 2 and 3 (and hence divisible by 6) – two independent events – is the product of the two separate probabilities

$$\frac{1}{2} \times \frac{1}{3} = \frac{1}{6}$$

- What would be the probability of tossing a head on a coin and a 6 on a die?
- Can you see why the probabilities of getting combined events is always less than of the separate events?
- Will the probabilities continue to decrease?
- What would be the probability of taking a sample which is a multiple of 3, 4 and 5?
- What would be the probability of taking a sample which is not a multiple of 3, 4 and 5?

Activity 6.15 This activity is similar to Activities 6.7 and 6.8. Multiply the scores on two dice and record the product. After 10 minutes of throwing, find the frequency of occurrence of the different products. Did you throw all the possible products? Now make a multiplication square for the numbers 1 to 6. Make a list, in order, of the number of times each product occurs in the table. How close were the experimental results to the results calculated from the table? From the table, calculate the probabilities of each product occurring.

Recognizing situations where estimates of probability can be based on equally likely outcomes and others where estimates must be based on statistical evidence.

We return now to the digit game introduced at the beginning of this chapter.

Activity 6.16 Collect 36 car numbers and ignore the letters – e.g. LGK 395 P is recorded as 395. (Avoid the school car park to prevent duplication!)

- Add the digits: 3 + 9 + 5 = 17.
- Add the digits again to get a single digit: 1 + 7 = 8.
- 'Add the digits' of each car number in your collection of 36, remembering to add a second time, when necessary, to reach a single digit answer. Allocate the results to sets 1 to 9, and find the totals for each digit 1 to 9.
- What can be said about the probability of obtaining a number which is a multiple of 9?
- What about any other of the digits?
- Will the probability be the same for each digit?

When this activity is introduced to pupils, they should collect as many car numbers as they can, since this activity provides good practice for quick addition. You may discover a secret about the quick addition of these digits – it involves 9.

Activity 6.17 Prepare a set of cards, numbered 1 to 20.

- What is the probability of drawing a square number from the set of cards?
- What is the probability of drawing a number which is not a square?
- What is the probability of drawing a prime number from the same set?

Knowing that if each of n events is assumed to be equally likely the probability of one event occurring is $1/n$.

The pupils have already carried out a number of activities in which they found, for example, that when throwing dice, the chance of throwing a specific number is $1/6$; the chance of not throwing this number is $5/6$. Ask them to look through the experiments they have completed and to list the outcomes.

Key Stage 3

Appreciating that the total sum of the probabilities of mutually exclusive events is 1 and that the probability of something happening is 1 minus the probability of it not happening.

This Programme of Study applies to manufactured goods such as electrical goods. For example, if the probability of a fluorescent tube failing is 0.05, then the probability of it not failing is 0.95 (1 – 0.05).

Teachers can suggest any other examples, and when they discuss this example with their pupils, the pupils may also be able to supply further examples from their parents.

Return to different experiments in probability from time to time, to keep the subject fresh in pupils' minds. It will be new to many of them and will give them much enjoyment.

Chapter 7

Cross-curricular Projects

INTRODUCTION

Throughout the National Curriculum reference is made, from time to time, to the importance of providing problems which can be developed across different subjects of the curriculum. We therefore include a number of interesting problems of this type, which have arisen and been solved at different levels.

We begin with an account of a project on 'Communications' which was used in an infant school with children of ages 5 to 7+, and then a similar cross-curricular project, 'The Post Office', which involved the children in a junior mixed and infant school in another area.

COMMUNICATIONS PROJECT (5–7 YEARS)

This topic was carried through with undiminished enthusiasm for a whole term. The Programmes of Study in mathematics, English and science for Key Stages 1 and 2 were all considered in the planning. Mathematics was well catered for.

The materials required included:

- used envelopes of assorted sizes, used stamps;
- a card for each child with their name, address and class number printed clearly;
- two postman's sacks of different sizes for each class;
- real money;
- (for the older pupils) individual number lines marked 0–100 (on the lines), cut from centimetre-squared paper, 1 metre long and 2 cm wide.

Key Stage 1 activities

At Level 1, the envelopes and stamps were used for sorting, comparing for size, counting and rearranging according to size (length, width and area).

Conservation of number This concept was reinforced when the children rearranged the envelopes in different ways – in a row, in a column, in a fan shape.

Counting Small stacks of envelopes were given to the children to count. They recorded the numbers and put the correct numeral on top of each stack – to be checked later by their teacher.

Number work Each child was given an envelope containing up to 10 stamps, which they counted. They were asked questions such as:

- How many stamps would be left if I took 2 away?
- What if I gave you 3 more?
- What if I doubled the number?
- What if I halved the number?

Estimation The children were helped to make sensible estimates of numbers (their first estimates were often wild guesses). For example, a small handful of envelopes was spread out fanwise, and the children were asked if they thought that there were more than 2 . . . 3 . . . 10 envelopes. They were encouraged to think about their guesses and gradually they became more proficient. The teachers, too, came to accept that in estimation an exact answer was not needed – the estimate should be 'near enough'.

Further Key Stage 1 activities

At this stage, the envelopes were used again. The children were asked to pick out one of the largest envelopes and to estimate how many handspans long it was. Each estimate was followed by counting to check.

Counting to 10 The children prepared the postman's bags with letters for delivery to other classes. At first, ten letters were put in each sack and postmen were chosen to deliver the post to four classes. When they returned to the 'sorting office', the postmen recorded what they had delivered to each class (e.g. 3, 3, 2 and 2 letters). The number of classes was varied but the number of letters was the same on any one day (to cover different combinations of 10).

Number work The name and address cards were used for arranging the house numbers in order (either ascending or descending). Small groups were given 10 or 12 cards and asked to find 'half that number', and later on, 'a quarter of that number'.

Sorting The children were given small sacks of used stamps to sort according to value. They were then encouraged to re-sort the stamps according to their own criteria.

More number work The stamps were also used to find differences between costs. Problem cards were provided. For example:

- Could I buy an 18p and a 24p stamp with a 50p coin?
- Would there be any change?
- If so, how much?

(The cards were illustrated to minimize difficulties in reading.)

More estimation The children were asked to estimate how many envelopes of a given size would cover the table. Each group chose a different size of envelope as a unit; the different answers prompted a discussion of the need for a standard unit of area.

It is clear that throughout this project the children were using numbers in recognizable, everyday situations and at the same time experiencing the concepts with obvious pleasure.

THE POST OFFICE PROJECT

Pupils at Key Stages 1 and 2 worked at this project for a whole term. Because the communications project involved children at Key Stage 1, for this project we focus attention instead on problems undertaken by pupils at Key Stages 2 and 3. There follows a summary of work covered at these stages.

Preparation work

Before they began the project each class prepared the Post Office for their own use. Some made a table into a counter at which two clerks served the 'public'. Others used plastic wire made into a screen to protect the clerks when working in the Post Office in the event of a 'hold-up'! They made tills to hold real money (to give change) and a drawer or a strong box to hold the stock of stamps – also made by the pupils.

Stamps, first and second class, were in sheets of 50, or in books of 4 or 10. Air letters were also made. Other pupils painted notices, such as POSITION CLOSED (or OPEN).

The first duty of the counter clerks was to check the total value of the stock and cash. This total was checked against the previous day's total in the account book. At the end of the session, the total value of stock and cash was again checked and entered in the account book.

Transactions in the Post Office gradually increased in difficulty.

Key Stage 2 activities

At Key Stage 2, pupils could purchase two or more items, offer cash, check the change and record the transaction. At this stage, the subtraction was always carried out by 'shopkeeper's addition' (the pupils' choice).

Introducing the concept of mass

The teacher introduced dial scales for weighing letters and packets. The pupils had already discovered from their local supermarkets that sugar, flour and other dry goods were sold in kilograms. The teacher then provided top-pan balance scales, large plastic bags and fasteners, and metal kilograms. She asked the groups to make up kilogram bags of different materials; sugar, flour, beans, water and Plasticine were used, and these were put on display. The pupils were astonished at the different sizes of the kilogram bags. The teacher asked them how the kilograms were the same. Jon said:

> 'They all have the same mass, one kilogram.'

'So, how are they different?', the teacher asked. Meryl said:

> 'They're different sizes.'

The teacher reminded them that there was another word they could use instead of 'size'. Jon volunteered the answer 'volume'. The teacher agreed and asked them to explain why the kilograms were of different volumes. Nirva suggested:

> 'Each bag is made of up different material. Metal is heavy, and flour is light.'

The teacher then explained that kilograms were units of mass and asked them to refer to 'finding the mass' (rather than weight) when using kilograms and grams. They all knew that 1 kilogram was 1000 grams 'because kilo means 1000'.

Reading the scales Before the pupils began to use dial scales for finding the masses of letters and packets, the teacher asked them to examine the scales to find out how many grams there were between successive graduations. Leslie said:

> 'There are 5 spaces between the 0 and 50 g mark, so each space between two successive marks is 10 g.'

Pricing The pupils then began to use the scales to find, and record, the mass of each letter and packet. One group made a poster showing the postage for each 'band' of mass.

	POSTAGE	
Mass not over	*First class*	*Second class*
60 g	25p	19p
100 g	38p	29p
150 g	47p	36p
200 g	57p	43p
250 g	67p	52p
300 g	77p	61p
350 g	88p	70p
400 g	£1.00	79p
450 g	£1.13	89p

and so on. At this stage, many pupils wrote letters of different lengths. The teachers encouraged them to make up problems on postage, to solve the problems and to enter these in a class book of problems. A favourite problem, by Indepal, was:

> How many sheets of notepaper could you use so that the postage on the completed letter was not greater than 19p?

At Key Stage 2, the activities were extended to include ParcelForce (post). Once more, a poster of the postage for parcels of different masses was made. This included parcels from 1 kg to 'not over' 30 kg, and postage from £2.50 to £7.80. Before the pupils began working with parcels of different masses, they made a block graph showing the postage due on parcels.

Julia had noticed that the postage (per kilogram) on a parcel of 30 kg was small compared with the postage on a 1-kg parcel. She said that the block graph did not seem to show this difference; she made a relationship graph, showing the mass along the horizontal axis and the postage per kilogram on the vertical axis.

Key Stage 3 activities

At Key Stage 3, some pupils began to investigate the mass of a single sheet of paper. The teacher provided two unopened packets of typing and duplicating paper, one of good quality, the other 'copy' paper. She had removed the two labels showing the mass of a square metre of the paper (e.g. 60 g per m^2). (When the pupils had finished their experiment, she let them compare their answers with the statements on the labels.)

Their teacher asked them first to estimate, then to carry out an experiment and finally to cut a piece of paper they thought would have a mass of 1 g. The teacher then asked the pupils how they planned to find the mass of a single sheet of the lowest-quality paper. One group began by using the letter scales (graduated in 10-g units to 1 kg). They tried to find the mass of a single sheet of paper. The pointer moved but not by 10 g. Laura suggested finding the mass of 10 sheets, and this time the pointer showed 45 g. Laura recorded that since 10 sheets had a mass of 45 g, 1 sheet would have a mass of 4.5 g. Jason thought they ought to check by finding the mass of 20 and 40 sheets. Eventually they recorded the mass of 20 sheets (90 g), 40 sheets (180 g), and 100 sheets (450 g).

The teacher then encouraged the pupils to try to think of another method for finding the mass of one sheet of paper (A4 size). The pupils were now about to use the higher quality paper. Pauline suggested finding out how many sheets of paper would have a mass of 100 g or 500 g. The pupils decided on 500 grams, all agreeing to join in the counting – 500 g is a lot of paper! Each of the five groups was given 100 g of paper to count. There were 11 sheets in each pack of 100 g. Pauline said:

> '11 times 9 g is 99 g, that's nearly 100 g. So each has a mass of nearly 9 g. So this paper has nearly twice the mass of the first.'

The pupils then tried to cut a gram of each kind of paper. Some folded one sheet carefully into eight equal pieces and cut one piece, saying it was not much more than 1 g. Laura measured the length of one side in centimetres (27 cm), divided this by 9, and cut a strip of paper 3 cm wide. She then cut a 6-cm strip from the copy paper, and checked by balancing the two pieces on top-pan balance scales.

MORE INVESTIGATIONS

The investigations which follow were undertaken in many different primary schools. Many of them originated in subjects other than mathematics but subsequently developed a substantial mathematical content. Some arose from questions asked by the teachers, others were raised by the pupils themselves.

Although the questions are divided into two sections (ages 8–10 years, and ages 10–13 years), many of the investigations could be undertaken at either level.

Cross-curricular investigations of this kind are strongly advocated in the National Curriculum. The list which follows may provide some useful starting-points. Almost all of these problems were solved by means of first-hand experience and experiments devised by the children (rather than solutions provided by teachers). We hope that you will be able to help your pupils to work in this way.

Investigations for 8- to 10-year-olds

Pulse rates 'My pulse beats billions of times faster than my heart'. How could you help an 8-year-old boy to check this statement?

Bird-table visitors This school regularly put out food for the birds, and were interested in the activity of the visitors to their bird table. The teacher asked them:

- How many birds visit the bird table during the ten minutes after we have put out the food?
- How many species of birds visit the bird table?

The teacher suggested that the children should look at the birds' beaks carefully and check from a book to help them to decide whether each bird was eating the food they would expect. She asked them to keep a record of the foods they put out and of their observations.

Growing plants from seed Having fed birdseed to the birds the question arose: Will all the mixed seeds in 'birdseed' grow? The children planted 2 or 3 of each kind of seed in soil in separate compartments of an egg box, or in separate plastic pots. They tried to identify each seed and to label it (or make a drawing). They remembered to include the date of sowing and of germination. The teacher challenged them:

- Can you find the percentage germination for different kinds of seed?
- Which conditions encourage growth?

This investigation extended to wild flower seeds collected by the children in the autumn and sown in the spring; again they found the percentage germination for each seed. (They were careful not to include rare flowers.)

Temperature How do our class temperature charts compare with those printed at the local weather station, in the local paper, and in the national newspapers? For this investigation the children included:

- classroom temperatures;
- temperatures outside the school building;
- soil temperatures at different depths;
- tap water indoors and outdoors.

From these data, many questions can be answered.

- What are the maximum and minimum outdoor temperatures during the month?
- What is the range of temperatures daily? for the month?
- How many days have the same midday temperature?

The children should be encouraged to make a note of interesting features of their temperature charts.

Which way does the wind blow? Some 8-year-olds made wind arrows to find the direction of the wind. They cut the wind arrow from corrugated cardboard. The following questions arose:

- How can we balance the arrow on a nail? (They used a ball-point cap inside one of the corrugations.)
- How can we find the arrow's point of balance?

The teacher suggested that the children should make a model see-saw, using a metre stick balanced on a wedge, and identical cubes as 'children'. They were then asked to investigate the relationship between the mass used and the distance from the fulcrum (point of balance).

Flower diversity This investigation started by considering how many petals a celandine has. The children counted and recorded the number of petals of at least 250 flowers. Questions were then raised.

- What is the most usual number of petals?
- If the number of petals had been the same for all the celandines in your sample, what would this number have been?
- How many kinds of yellow flowers can you find with 4, with 5, and with 6 petals?

Popular pets The children investigated which is the most popular pet in their class.

- Is the result the same in any other classes in the school?
- Has anyone in your class an unusual pet?

The teacher then asked them to find out how much it costs to feed various pets for a week, a month (4 weeks), and a year (52 weeks).

Box modelling This activity involves gluing boxes together, cutting paper to fit each face and pasting to cover the model – and finally painting the whole to make it strong and aesthetically satisfying. It is a very fruitful source of mathematical ideas. The problems which follow have arisen in work of this kind.

- How do you recognize a cube? a square? a rectangle?
- What shapes are the starting-points for making: a cylinder? a cube? a cone?
- How can you alter the shape of a cone?

Shapes Different shapes are all around us. The children can investigate the shapes of buildings and objects they see in a street.

- Compare your results when you visit a street in another locality.
- Sketch the buildings you see from the top floor of a high building.
- Can you name all the shapes you can see?

How tall was Goliath? (6 cubits and a span) Some 10-year-olds experimented with the cubits and spans of various children and of obliging adults in order to find Goliath's actual height. They could then investigate the range of the results in their own school.

Tessellations A lesson on Roman history led to a visit to a Roman villa. The children asked why the Romans used square tesserae (small tiles) for the mosaics.

- Which other regular shapes could be used?

Experiment to see what can be discovered.

Distances to school For this investigation, the children need to use a local map with a scale.

- Who lives farthest from and who lives nearest to school?
- How many children live more than a kilometre from school?
- Compare this with the number of children who live less than a kilometre from school.
- How much shorter is the distance 'as the crow flies' than your actual journey to school?

Speed Rolling balls and model cars down a slope in the playground prompted the following questions:

- Which vehicle reaches the bottom of the slope first?
- Is it the same vehicle every time?
- Which one rolls farthest along the flat ground?
- Was this vehicle the fastest?
- How did you make sure that the starting conditions were the same for each vehicle?

Animal feeding costs daily at a West Country zoo In Figure 7.1 all the costs are given in pounds. (You may like to check the costs when you next visit your local zoo.)

- Which is the most expensive animal to feed?
- How much does it cost to feed this animal for a week (7 days)?
- Which is the least expensive animal to feed?
- What is the difference between the daily costs of food for these two animals?
- Make a list of the daily cost of food for animals when this is more than £1.
- Arrange the costs in order from the most to the least expensive.
- Find the weekly total for each of these animals, and the grand total.
- Find out the type of food each animal eats. Do you think that the cost depends on the type of food the animal eats or on its size? Or on both?
- How many days would it take for a camel's food to cost more than an elephant's food for one day?

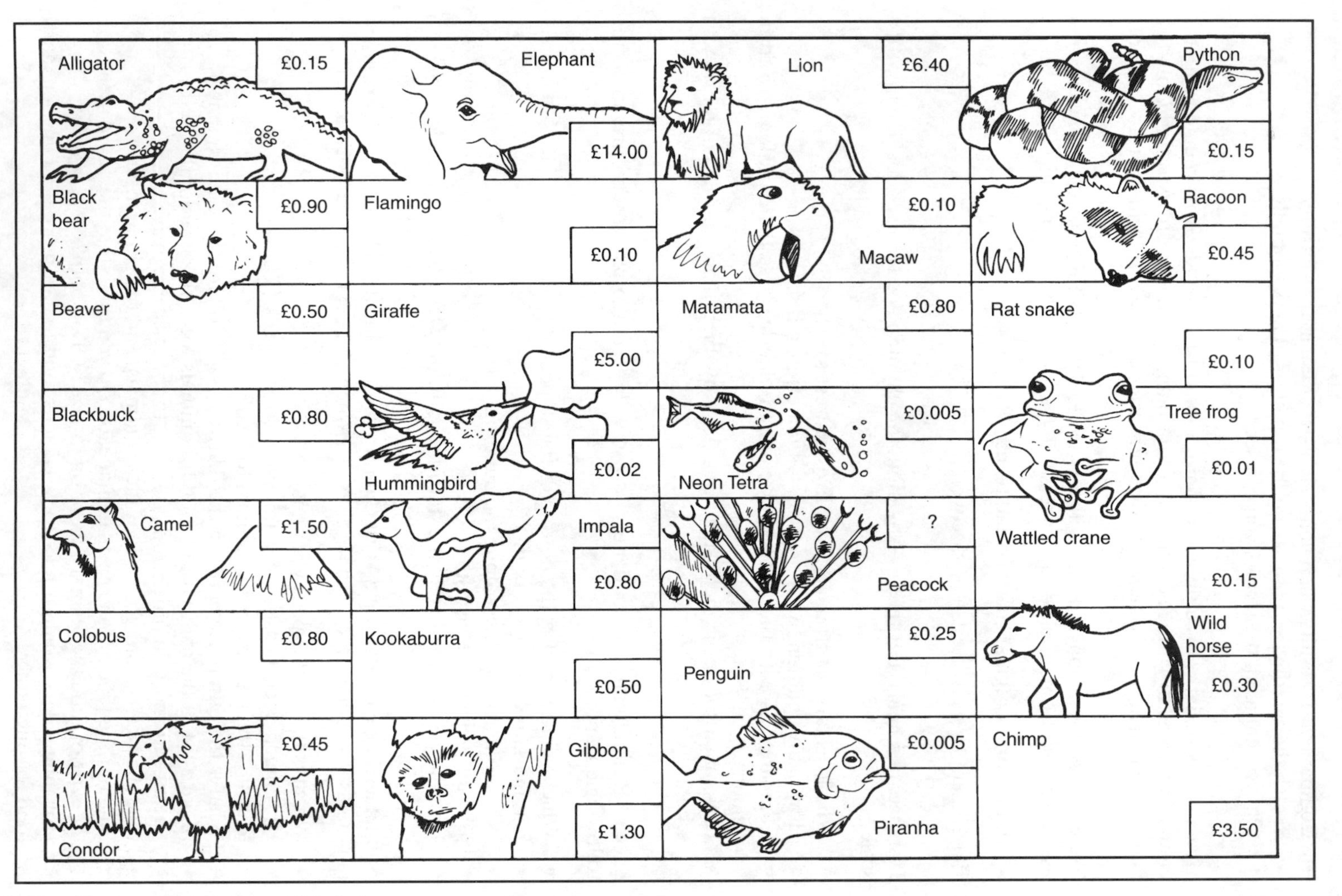

Figure 7.1 *Animal feeding costs daily in £*

Investigations for 10- to 13-year-olds

Circular measures How large is a ball?

- Find the diameter of a large ball in as many ways as you can.
- Compare your methods.
- Measure other properties of the ball.
- Find a ball of half the original diameter and take and record its measurements.
- Compare measurements for the two balls.
- Obtain a set of balls made of the same material but of different sizes and compare their various measurements.

Length, area, volume, pressure and strength Use identical cubes or cuboids to make some multiple scale models (2-scale, 3-scale, etc.) to investigate these problems.

- Why do large animals such as the elephant seem to have outsize feet?

Identify the variables you need to consider: length? surface area? volume? mass? Any others?

- What is the average pressure on an elephant's foot?

This question was solved by 10-year-olds by obtaining a footprint and information about the mass of an elephant from their nearest zoo.

- Obtain the necessary statistics for two of the lightest children and two of the heaviest children in your class. (First make sure that they would not be embarrassed by this enquiry!)
- Comment on your results.
- Obtain similar statistics for any available animals.
- Compare your results.

The stilt and the flamingo are both long-legged birds and are of approximately the same shape. The corresponding leg lengths, as measured by Gilbert White of Selborne (*c.* 1770), were 7″ and 28″. This made Gilbert White expect that the weights of the two birds would be in the same ratio as the lengths of their legs. He found his estimate to be far out.

- Experiment with enlarging identical unit cubes or cuboids to find out where Gilbert White made a mistake.

After a BBC programme on the blue whale some 10-year-olds made some scale drawings of blue whales and classrooms. They found that it would take 3½ classrooms to accommodate the length of an adult blue whale, whereas a baby one would fit comfortably into one classroom. They next calculated that the ratio of the lengths of adult to baby whale was 4:1 whereas the ratio of their masses was 15:1. They were so puzzled by this that they calculated the corresponding ratios for human beings: lengths 3.6:1 and weights 14:1.

- Can you account for these results? Use identical cubes for this investigation.

Soil studies This investigation finds the range of temperatures within the soil:

- at different depths;
- in different locations, e.g. in full sun, under a tree, etc.

Set aside equal masses of each soil sample. First, estimate the percentage mass of water which each sample holds, and then check by experiment:

- What is the range of these percentages?
- How did you make sure that each sample was thoroughly dry?

Next we make a model to help us to understand the porosity of soil. Fill a graduated container with identical marbles, and estimate the fraction of the container which is taken up by air.

- Check the result by experiment.
- Discuss your methods.
- Was the volume of air about one half of the volume of the container?
- Does this surprise you?
- If you use larger marbles, or marbles of mixed sizes, do you get a different result?
- Is this problem related to the porosity of different soil samples? Investigate this problem.

Insect life This investigation concentrates on insect life in a small area of the school garden. The children record location, type of insect, its 'home', its 'work' and its food.

- How far did it travel from its base?
- Can you find its speed?
- Can you estimate the size of the community?
- Record any interesting observations you make.
- Compare your observations with those in a reference book.
- Examine the insect under a lens and make a drawing.

Plant life As in the previous investigation, the children record the number of different plants in a selected area. They can use a square of wire or string, 2 m in perimeter, to enclose the plants in a $1/4$ m^2, chosen at random. For example, they throw a stone on to the patch they are studying, spread out the square with the stone at its centre and record the names of all the different plants enclosed. This can be repeated several times.

- Make a map of the areas covered, showing the plant population.
- Were your samples representative?

When you try this activity with children, a preliminary survey of the plants in the selected area may be advisable. Suggest that the pupils make a list of all plants. Better still, provide pictures which they can use to identify the plants listed. They can also make comparisons of plants in different areas.

Trends This activity introduces an important statistical concept: correlation.

- Measure the length and width of leaves from a branch or twig.
- Make a scattergram of these measurements.

- What degree of correlation does the scattergram show between the varying lengths and corresponding widths of the leaves?
- Comment on 'the line of best fit'.

Weather studies In England, we have many rainy days, so we consider rainfall first! Collect rainfall daily in a cylindrical tin with a tightly fitting funnel, transferring it to a glass or strong plastic bottle of diameter 3 – 4 cm. Seven identical clear bottles will be needed, then the children will be able to compare the amount of rainfall which falls daily for a week at a time.

- Describe how you calibrated the bottles in mm.
- How do your observations compare with those recorded at the local meteorological station and with weather reports in the local newspaper?

If the weather is really bad, aim to collect hailstones!

- How large is the largest hailstone you have found?
- Cut a large hailstone in half with a heated blade. How do you think the layers were formed?

If the sun shines, we can concentrate on recording sunshine. For this experiment it is best to choose a day when the sun is expected to shine all day long.

Use a cylindrical tin about 10 cm in diameter and 20 cm high. Punch a hole about 2 – 3 mm in diameter half-way up the tin. Line the tin with photographic paper, leaving the hole uncovered. Place the tin with the hole facing the sun. Make a note of your starting and finishing times (09.00 to 16.00?).

- How many hours 'long' is your 'sun line'?
- How long is one hour of sun?

You can 'fix' the sun line by washing the paper with water as soon as you remove it from the tin.

- What do you notice about the shape of the sun line?

If you stand the tin upright, the sun line will be curved.

- What will you have to do to obtain a straight sun line on the photographic paper?

Yes – you need to tilt the tin, propping it on something heavy such as a brick or stone. Begin by fixing the tin at 45° to the horizontal.

- Is the sun line less curved?

If so, the next day continue to increase the slope of the tin, until the sun line is horizontal. Of course, you can have two or more tins in the sun, sloping at different angles. (The angle of slope required is the angle of latitude – a useful example for your pupils to discover.)

Anemometer This is a good example for a windy day. Use two identical pieces of light wood; fix them in the form of a cross and pivot them at the centre, drilling a hole to fit a biro cap; fix this firmly in the hole. Fit four identical plastic cups (one of a different colour) on their sides to the free ends of the cross, so that they catch the wind and cause the cross to rotate. Balance the anemometer on a long nail or on a skewer. Support this in a bamboo cane.

- Does the cross revolve smoothly?
- Can you count the number of revolutions made in 1 minute? 5 minutes?
- At this rate, how many revolutions will be completed in an hour?
- How far does each cup move in one revolution? Calculate this from the circumference of the cross using $C = 2\pi r$.
- At this speed, how far does each cup travel in one hour?
- Compare your results with those in local newspapers.

Use the anemometer on days with different wind speeds.

Flight Arrows were made by pupils studying the exploits of Robin Hood. Some arrows were more satisfactory than others. They asked:

- Is this anything to do with the point of balance?
- The materials of which the arrows were made?
- The length?
- The thickness?

Investigate these problems raised by the pupils in their attempt to obtain the longest shot.

Great circles For this investigation you need a globe and a length of thin string.

- How far is it, by air, from London to Trinidad?

Explain how you measured the distance from London to Port-of-Spain. If you pulled the string tight to find this distance, you will have found the 'great circle' distance between London and Port-of-Spain.

- Can you think what 'great circle' must mean?
- Which other places on the globe are as far from London as Trinidad is?
- What can you discover about the great circle distances from London to Singapore? Moscow? Tokyo? Any other place which interests you?
- Investigate how long the journeys would take by air or by any other method of travel.

What shape is a bee's cell?

- Why do you think bees use this shape for their honeycombs?

One theory is that bees make circular cells and that these are pressed into hexagons. Is this feasible? Investigate other possible regular shapes which bees could use.

How does a kaleidoscope work?

- What is the effect of changing the angle between the mirrors?

Investigate the relationship between the angle made by two hinged mirrors and the total number of objects (and reflections) seen.

- Make a chart of your results.
- Do you recognize the relationship between the angle and the total number of repetitions?

Data handling This investigation has a historical flavour. Collect information from the parish records of two neighbouring villages and make any comparisons which interest you. (Permission must be obtained.) In one area, trades and crafts of 100 years ago were studied and the times of their disappearance were noted. Other characteristics studied were:

- occupations of newly-weds
- the number of occupants per house
- ages at death
- popularity of first names

Another useful source of information is *The Northumberland Household Book*. It is interesting to make comparisons with present-day figures, e.g.

- recipes
- cost of food per head per week
- standard measures

Also, looking at old county maps, comparing these with contemporary Ordnance Survey maps, and comparing modes and times of travel, past and present, can be an enjoyable and rewarding experience.

Strength of structures Before you give this activity to your pupils, take them to look at a girder bridge (or show them photographs). Ask them what they notice about the unit (basic) shapes. They may comment on the numerous identical triangles to be seen. Back in the classroom, you will need rigid strips of two lengths, e.g. 10 cm and 15 cm, and push-through paper fasteners. The strips can be made of metal (e.g. Junior Engineer or Meccano), wood or stiff card. Ask the children to use the strips and fasteners to investigate the rigidity of different regular shapes. You may like to start them with the problem:

> Why do you think the triangle is used so often in metal structures, for instance, roofs?

At first, let the pupils experiment freely – but make sure that they record all their results. After ten minutes, ask them to name their variables. They may suggest:

- number of sides
- shape
- size of angles
- number of struts (or 'stiffeners')
- number of triangles

Ask them to make an ordered table:

Number of sides	2?	3	4	5	6	7	8
Number of struts		0	1	2			
Number of triangles		1	2				
Sum of angles		180°	360°				
Each angle		60°	90°				

Ask them to complete the table as far as 8 sides, and then to describe each number pattern. They should then make graphs of the relationships, and describe these.

Point of balance This activity seeks to find the point of balance (i.e. centre of gravity) for a number of different shapes.

Cut a variety of two-dimensional shapes from thick card: square, rectangle, circle, semicircle, trapezium, parallelogram, irregular quadrilateral, various polygons.

Take the parallelogram first and poke a small hole near one edge; tie a piece of thin string through this hole and hold the shape by the string. While still holding the string, ask your partner to place a ruler against the card, continuing the line of the string down, across the shape. Mark this extended line on the card. Check that the direction is correct. Then, poke a second hole, again near the perimeter, obtain a second vertical and mark its direction. You now have two lines in two different directions. If they do not meet, extend them until they do. They should meet at the point of balance of the shape. Use the head of a roofing nail to pierce the card at the balance point, to check that you have indeed found the right point.

Repeat for the other shapes, and label each point of balance clearly. Another name for this point is the centre of gravity. Try to find the centre of gravity of some three-dimensional shapes: cuboid, cylinder, cone.

Another investigation in the same vein involves balancing a square wooden board on a large ball. Add structural materials (preferably Dienes Multibase Arithmetic Blocks) until the board is horizontal again. How many variations can you find? Invent a game to give practice in restoring balance when one mass is removed.

Elasticity Wire springs led some 11-year-olds to ask:

> Is there a relationship between the mass pulling on the spring and the extension it causes?

This experiment can also be tried with elastic bands. Investigate this problem.

Curves Find a good reference book such as Lockwood's *A Book of Curves* (published by CUP), and try out the various methods suggested for making spirals in two and three dimensions.

- What do you notice?

Cut a sector of a circle and draw a series of lines parallel to one straight edge. Fold the sector to make a cone with several thicknesses.

- What do you notice?

This problem arose from a study of climbing plants and shrubs.

- Make lists of those which climb in a clockwise direction and those which climb in an anti-clockwise direction.
- Do plants always climb in the same direction? How do they secure their climb?

Estimating heights For this experiment, we need a tree and some sunshine. First, choose a shady tree. Take any measurements which interest you. One rough-and-ready method of finding the height of the tree is as follows.

- Take an unsharpened pencil.
- Hold it vertically and at arm's length.
- Walk towards or away from the tree until the pencil appears just to cover the height of the tree.
- Swing your pencil through a right angle until it seems to lie along the ground, starting at the foot of the tree.

- Now ask a friend to walk away from the foot of the tree until you tell him to stop – when he appears to reach the free end of the pencil.
- Measure the distance of your friend from the foot of the tree.

Why should this distance give you the height of the tree?

For a more reliable method of finding tree heights, you can use the shadow method: compare the lengths of the shadows of a metre-stick and of the tree.

Other measurements which might interest you are:

- the approximate area of the shade cast by the tree;
- a comparison of the ratio of the lengths of the smallest and largest leaves on a long twig, and of the ratio of the areas of these leaves.

The Golden Section and the Golden Rectangle

Figure 7.2 illustrates this concept.

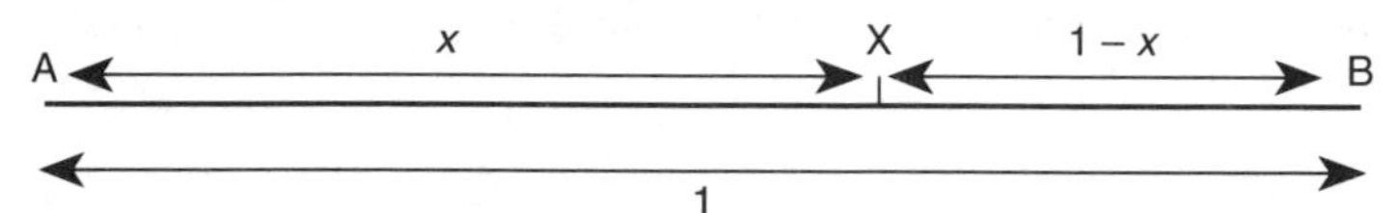

Figure 7.2 *Golden Section*

The point of Golden Section of a line AB is at X which divides the line so that the ratio of the larger part to the smaller is the same as the ratio of the whole line to the larger part.

$$AX/XB = AB/AX$$
$$\Rightarrow AX^2 = AB.XB$$

Suppose the length of AB is 1 and the length of AX is x, then

$$x^2 = 1(1 - x)$$
$$\Rightarrow x^2 + x - 1 = 0$$
$$XB = 1 - x$$

The solutions of this equation are:

$$x = \frac{-1\ (+\text{ or } -)\ \sqrt{1 + 4}}{2}$$

or

$$x = \frac{(\sqrt{5} - 1)}{2}$$

which is

$$x = \pm 0.618$$

but since x must be positive we take $x = 0.618$.

In the Middle Ages the focal point of paintings was often located at the Golden Section. If you investigate reproductions of various Italian masterpieces, you will find that the pictures are themselves often enclosed in a Golden Rectangle, with sides in the ratio of the Golden Section – 1:0.618.

In fact, the Golden Rectangle has always been considered a pleasing shape. This can be investigated by asking pupils to choose the rectangle whose shape they prefer from a variety with sides in different ratios: 2:1, 2.5:1, 3:1, 1:0.618.

Fibonacci series

The Fibonacci series was named after the Italian who discovered it in about 1200 AD. Find and describe the pattern of this series:

1 1 2 3 5 . . .

Continue the series for another ten terms. The Fibonacci series is closely associated with the Golden Section, and often occurs in nature: for example, the number of seeds in intersecting spirals in a sunflower head, e.g. 34 and 21; and in comparable spirals in fir cones, e.g. 5 and 8, 8 and 13.

The position of the Golden Section in a line can also be found from a consideration of the ratios of adjacent terms of the Fibonacci series:

Fibonacci series	1	1	2	3	5	8	13	21	34
Ratio of adjacent terms	1:1	1:2	2:3	3:5	5:8	8:13	13:21	21:34	34:55
$\geqslant 0.618$	1		0.667		0.625		0.619		0.618
$\leqslant 0.618$		0.5		0.6		0.615		0.618	

Why do you think that we have arranged the ratios in alternate rows? Did you notice that in one row the initial ratios are greater than the Golden Section and diminish towards it; in the other row they tend to it from below?

Golden Rectangles can be used to make a spiral (see Lockwood's *Book of Curves*). The diagonals of regular pentagons intersect at points of Golden Section (see also Lockwood).

Chapter 8

Calculators and Computers

INTRODUCTION

The extent of the use of calculators in primary schools varies a great deal. However, it is becoming increasingly recognized that calculators are valuable tools, not only in junior classes but also in infant classes, provided that they are used imaginatively. The Calculator Aware Number project (CAN) introduced calculators into Year 1 classes; calculators were made available to all children at all times, each child having his or her own calculator. The project was successful and demonstrated that children were able to handle numbers flexibly. At an earlier age than expected, these children:

- were good at mental arithmetic;
- understood place value;
- were familiar with large numbers;
- grasped the concept of negative numbers;
- accepted the concept of decimal fractions;
- were able to investigate number patterns.

The use and ownership of a calculator had a motivating effect and removed anxiety about incorrect answers – it was the calculator's fault! And you could always 'clear' it. However, the recording of work using calculators gives few details of the calculation, so it may be more difficult for teachers to identify errors.

Calculators, like computers, have an attractive magical quality for young children and this aspect needs to be capitalized. At the present time, however, children are introduced to calculators at different stages in their mathematical development. The National Curriculum does not mention the use of calculators until Level 3, and then mostly in the context of algebra. This conservative approach is comparable with the approach to computers and may well be due to the reluctance to assume a certain level of resourcing in schools. However, in many cases, the use of the calculator is not intended to satisfy National Curriculum requirements *per se*, but to facilitate the study of mathematics in general. The introduction and progression, particularly in the early stages, will therefore vary.

In the activities which follow many of the questions are not as open-ended as those one would use in class, but they are set out in this way to indicate the progression of ideas. A fairly early starting-point is assumed, when children can recognize the digits 0 to 9 and count to 10 reliably. Not everyone would agree, but for a large number of classroom teachers this is an easy starting-point.

What type of calculator is best?

The actual make of calculator is not so important. However, it is easier for the teacher if, to start with, all the children have identical calculators. Particular keys and the way they work can then be discussed without confusion. (It is also possible to cover with a mask those keys which the children are not to use.) Children should be first introduced to a simple calculator, i.e. one with numbers and simple operations only. It is important that, while experimenting with calculators, children should still use other apparatus, such as Unifix, Multilink, Cuisenaire rods, etc., all of which should be readily available.

When using activities described here and in other sources, it is essential to check first that each does work in the same precise way on your calculator version. Not all calculators produce the same result from a given key sequence, nor do they all give access to certain useful functions, such as the constant factor function, in exactly the same way. Therefore it may be necessary to refer to the instruction book for your calculator or to ask someone!

CALCULATOR ACTIVITIES

Play is always an important starting point with young children (in fact children of any age, including teachers!).

The calculator design

Before a calculator can be used properly all the keys that can be pressed, and the display, should be understood. So, as a starting point, ask the children to look at the numbers on the keyboard, and see if they recognize them.

- How many numbers fit on the display?
- Can they put the entire keyboard set of numbers into the display?
- Which key do they use to turn the calculator on and off, to empty the display and to start afresh?

It is also a good idea to spend a little time looking at the actual structure of the numbers on the display.

- Ask the children to make the numbers by sticking matchsticks onto card.
- Investigate the number of matchsticks needed for each number.
- Given a number of matchsticks, what is the biggest (or smallest) number that can be made?

These activities could contribute to the Programmes of Study for Number at Key Stage 1.

Looking at the + key – what does it do?

Here * is used to mean 'press the key', and we concentrate on what happens when we press the + key and the = key. As a first exercise, follow this sequence:

* 1 * + * 1 * = * = * =

- What is happening?
- Can you reach 12?
- How many times did you press the = key?
- Does the number always get bigger?
- Will the calculator count for you?

On some calculators, the sequence needed is

* 1 * + * 1 * + * 1 * + . . .

So it is important to check exactly how the calculators available actually do work.

- Can you use this method to count how many steps you take to reach –?
- Can you press 1 and use the + key and other number keys to reach to any given target number?

Discussion is important at this stage to establish firmly in the children's minds what the + key does and what pressing = does.

Activity 8.1 Extend this activity to lead on to work with odd and even numbers. Try

* 2 * + * 2 * = * = * =

and

* 1 * + * 2 * = * = * =

or their equivalents, and discuss the numbers obtained.

- Can the sequence be continued?
- Which pattern would you need to use to get to 37 or to 100?
- Is there only one way for each?

Activity 8.2 Use some numbers and the + and = keys to get from one number up to another.

- How many keys did you press to get from 3 to 23?
- Could it be done by pressing fewer keys?

Activity 8.3 Ask the children to use some numbers and the + and = keys to reach target numbers, recording their work using formats shown in Figure 8.1.

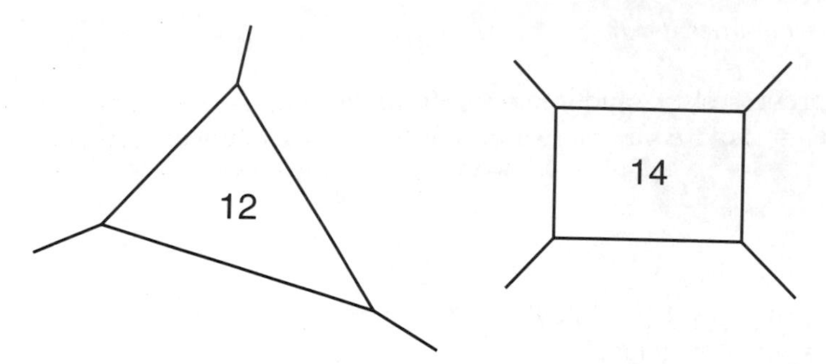

Figure 8.1

These activities could contribute to statements in Programme of Study 2.

Looking at the − key

Activity 8.4 As with the activities for the + key, the first step is to consider counting down, e.g. the sequence

∗ 20 ∗ − ∗ 1 ∗ = ∗ =

or its equivalent for your calculator. This should provide an opportunity to discuss the use of the − key.

Similarly, the sequences

∗ 20 ∗ − ∗ 2 ∗ = ∗ =
∗ 23 ∗ − ∗ 2 ∗ = ∗ =

will provide starting-points for odd/even number sequences. Try using an odd number in the sequence, e.g. 3

∗ 25 ∗ − ∗ 3 ∗ = ∗ = ∗ =

giving a series of alternate odd and even numbers. Some of these activities, similar to those for the + key, can be used to lead to worksheets and investigations.

Negative numbers

The calculator also provides a convenient way into negative numbers, since it may be appropriate here to examine what happens if you go on using the − key past 0. In the display the minus sign can be rather unobtrusive and may need to be pointed out to the children.

Number line The calculator will produce large numbers. The children can make a number line to record these numbers on it. This should be continued to include large numbers which will need to be identified, such as significant year numbers, e.g. 1995, 1945, 1066, etc. and later, an extension of this number line past 0 will help to consolidate the concept of negative numbers.

These activities contribute to statements in Programmes of Study for Algebra and Number. The Levels need to be assessed taking into consideration the mathematical knowledge and understanding of the individual pupil. We need to remember that the calculator is only a tool for the mathematics.

Looking at patterns

We have already considered odd and even numbers and some children may have gone further. If given more examples of continuous addition, some children will recognize their multiplication tables! These children could then be asked to explore the use of the × key and could see how quickly the numbers grow.

These activities cover statements in Programmes of Study for Number and Algebra.

Place value

Activity 8.5 Place value can be explored by looking at the patterns obtained by the continuous addition or subtraction of 1, 10 or 100, or perhaps even adding or subtracting 1000, e.g. by using the sequence

∗ 1 ∗ + ∗ 10 ∗ = ∗ = ∗ =

Discuss the numbers obtained and provide similar activities until the children can predict the patterns.

Activity 8.6 Start by keying in a three-digit number and use the continuous addition (or subtraction) of 10 several times, and similarly of 100.

Use different starting points and different targets, and discuss with the children what they are doing. It is valuable to write down some of the sequences so that the children can examine them.

These activities contribute to statements in Programmes of Study in Number and Algebra.

Investigating the + , − , × and ÷ keys

Activity 8.7 Quite soon, children can be encouraged to try some 'sums' on their calculator. Ask them to enter these sequences and to record their results.

∗ 3 ∗ + ∗ 2 ∗ =
∗ 5 ∗ − ∗ 2 ∗ =
∗ 6 ∗ + ∗ 3 ∗ =
∗ 6 ∗ − ∗ 3 ∗ =
∗ 6 ∗ × ∗ 3 ∗ =
∗ 6 ∗ ÷ ∗ 3 ∗ =

They should be given many more similar sequences, or sums, to do, so that they soon recognize the use of the four operation keys, + , − , × and ÷ .

Activity 8.8 They can then try longer sequences which will reinforce their understanding of multiplication as repeated addition, and division as repeated subtraction.

∗ 0 ∗ + ∗ 3 ∗ + ∗ 3 ∗ + ∗ 3 ∗ + ∗ 3 ∗ =
∗ 3 ∗ × ∗ 4 ∗ =
∗ 12 ∗ − ∗ 3 ∗ − ∗ 3 ∗ − ∗ 3 ∗ − ∗ 3 ∗ =
∗ 12 ∗ ÷ ∗ 4 ∗ =

* 0 * + * 3 * + * 3 * + * 3 * + * 3 * =
* 4 * × * 3 * =

It is important that they remember to clear the display before each sum.

Decimal point

Activity 8.9 The use of the ÷ key is bound to lead (with an appropriate choice of sequence!) to the appearance of a decimal point. This may already be familiar from its use with money. Investigation of the function of the ÷ key (using small numbers, combined with the use of number apparatus) will help children to understand that decimal fractions are numbers in between the whole numbers, e.g.

* 10 * ÷ * 3 * =

These activities contribute to the requirements of Programmes of Study for Number and Algebra.

Estimation and approximation

Activity 8.10 It is good practice to encourage the children to estimate the approximate size of the answer expected, before they begin each calculation. The checking of an answer to ensure that it is sensible is a very good habit to instil. It also develops a 'feeling' for the size of numbers, e.g.

* 119 *×* 8 * =

i.e. the answer will lie between 800 (that is, 100 × 8) and 1200 (that is, 120 × 10).

Many of the calculator games to be found in such resources as Calculator Workshop from the Mathematical Association are helpful in promoting these skills. These activities cover statements in Programmes of Study for Number and Algebra.

Calculators as a tool in investigations

Activity 8.11 Because the calculator can give the child the answers quickly it is a good tool in investigations. It makes the use of real data a possibility once the child has grasped the operations involved. To start with, they can use the calculator to investigate, for example, the addition of odd and even numbers, doubles, halves, squares, cubes, fractions such as $1/2$, $1/3$, $1/5$, $1/10$, and patterns such as

9 × 4
99 × 4
999 × 4

For this last exercise, there will be a maximum number which can be calculated. But the children will discover that they can carry on and predict the next calculation that the machine can *not* do. (The machine will run out of space before they do!)

'Trial and improvement' methods

Activity 8.12 As an example, ask the children to find x when $x^2 = 12$ (without using the square root function key, if there is one!). The children might start by suggesting that x must be a number between 3 and 4. Someone suggests trying 3.5. The calculator gives 3.5×3.5 as 12.25. Trying 3.45 is the next offer. The children quickly discover that $3.45 \times 3.45 = 11.90$ – much nearer. They then continue this process until they are nearer still to 12. At some stage, the decision is taken that 'this is close enough' – a good time to introduce ideas of 'significant figures' and to discuss approximation.

Activity 8.13 The square of x is successively: 10, 20, 30, 40, 50. Find x each time to 3 significant figures.

Activity 8.14 The sum of two numbers is 12 and their product is 34. Use trial and improvement methods to find the two numbers.

Other button functions

The calculator will probably have a suitable 'clear entry' (C) key which should be explored by the children. Even basic calculators, used in the early years, may also have a square root key which can be useful, and is best introduced when it is needed. It is essential that the children should be allowed to explore *all* the keys on their calculator in due course.

In primary schools, from this point onwards, the calculator should be an established tool of the mathematics curriculum. However, as the children progress through the school they may be using more complex calculators. They need to be taught and allowed to investigate the full range of functions available on their calculators. The instruction book will list the functions available. On children's own calculators, the notation may be different and brackets, for example, may need to be handled differently.

CALCULATORS IN GENERAL USE IN SCHOOLS

Calculators are useful in the mathematics curriculum:

- to facilitate the use of real data;
- to explore numbers;
- to explore patterns in numbers;
- to reinforce place value;
- to introduce or consolidate concepts such as negative numbers, decimal fractions, vulgar fractions and division, the four operations;
- to facilitate the use of large numbers;
- to improve facility in mental arithmetic;
- to make calculations speedier when appropriate;
- to check calculations.

Summary: where calculators can be of use

Programmes of Study for Using and Applying Mathematics
The calculator serves as an aid to investigation and real-life problems.

- Level 1 The calculator acts as apparatus to facilitate tasks.
- Level 3 It is used in checking results.
- Level 3 It is used for investigating and for testing predictions.
- Level 4 It is used for testing solutions or definitions.

Programme of Study for Number and Algebra
The calculator is used in calculations at any level when handling numbers or operations.

- Level 2 Because operations performed on a calculator are quick and simple, many number facts are learned, e.g. halving and doubling numbers becomes well established.
- Level 3 The calculator can be used for arranging numbers in order of magnitude when multiplying or dividing. It emphasizes the need for estimating as part of the technique of checking results. Reading calculator displays helps children to read displays of both time and money.
- Level 4 Practice in estimating and the subsequent calculations help pupils to understand underlying concepts. Place value can be established early. The relation between fractions and division is soon understood.
- Level 5 Calculators are used for finding fractions and are essential for 'trial and improvement' methods. This is one section that needs calculators as a tool.
- Levels 1 and 2 Calculators can be used for making and exploring repeating number patterns.
- Level 3 Calculators facilitate using patterns for mental calculations.
- Level 4 They are used to assist in investigating number patterns.
- Level 5 They are used to generate number patterns on instruction.
- Level 6 They are used to explore number patterns.

Programme of Study for Shape, Space and Measures
The use of calculators here is not so obvious, but children often use calculators when working with LOGO, for example,

- Level 4 Calculators are used when constructing shapes.
- Level 5 Calculators are used when calculating volumes, areas, ratios, gradients, and in trigonometry.

Programme of Study for Handling Data

- Level 2 Calculators are used as a counting and recording device. Subsequently they can be used to manipulate data as required: e.g. for finding averages, percentages, etc. The use of the calculator is often vital if 'live' data are to be used.

USE OF COMPUTERS IN MATHEMATICS

Introduction

All pupils should have access to computers in the classroom, and these should be used, not only in all the core subjects, but in the other curricular subjects as well. The use of Information Technology is specified in the National Curriculum and the use of computers is specified in each area of the National Curriculum, so computers need to be used as a tool within other areas of the curriculum. Several uses of computers are indicated within the mathematics curriculum; one of the main uses suggested is that of LOGO as a programming language. LOGO is treated as a special case in Chapter 9.

In mathematics, the role of computers was recognized quite early on. The proliferation of uses for computers has meant that its original focus, mathematics, has been neglected to some extent. However, many of the original programs are still useful; for example, Microsmile, a product of the Smile project (available on the *BBC Master, A3000* and *Nimbus* machines), which includes 59 programs, is well worth considering for use, and others are still being released. These programs cover various aspects of mathematics and use different learning strategies.

Computers can be used for different types of work, from investigations to practice of skills. The programs can be used with all ability ranges, since they can easily be tailored to individual needs, and slower pupils, particularly, enjoy working with computers because they are not restricted by time and can usually have a second or third attempt at a problem before having to ask for help. For pupils of a higher ability range, investigations can be chosen which allow each to reach his or her full potential.

Adventure games

Many adventure games have a mathematical content: either mathematical problems have to be solved in order to progress to the next part of the adventure, or others are played on a $4 \times 4 \times 4$ grid, for example, and the children have to map their way through the spaces to avoid being trapped.

Generally, ideas such as triangular numbers, Fibonacci and other sequences, codes, coordinates, etc. are used. Some games for younger children test mathematical skills in terms of measures, money, space; for example, they have to answer:

- How many cakes, at 15p each, can you buy if you have 50p?
- Can you reach a shelf 2 metres high?

Some adventures are based on data handling. For example, in *ADVENTURE OF SORTS*, the members of the group enter all their personal data such as height, mass, reach, etc. and the adventure is then tailored to the group's data. They have to select the individual who fits the data prescribed for each part of the adventure.

Data handling

There is now a greater choice of data handling programs (data base packages) for schools to use, although one of the main deciding factors will be the make of machine it is to be run on.

DATA SHOW is available on the BBC and the A3000, and is a very good starting program for young children. It leads to the *MESU* data handling pack on the BBC. Other programs are available such as: *KEY*, *DATA SWEET*, *PIN POINT* (junior), *PIN POINT* and *DATA KING*. Each school will need to select perhaps two of these to cover their own range of age and ability.

On the Nimbus *OUR FACTS* is available, and also *COUNTER* (very good for the youngest children) and *CLIPBOARD*. The new *WINDOW BOX* machine has *STARTING GRID* at four levels – the top one being the adult version of *EXCEL* commonly used commercially. *GRASS* and *GRASSHOPPER* have been used a great deal on Nimbus machines.

Consolidation

For many children, work on the computer is a good way of consolidating a concept or a skill. For example, *MINIMAX* from the Microsmile programs gives children five random numbers to fit into a framework. They choose the operation (addition, subtraction, or multiplication) and whether to aim at a large (or small) result. The challenge is to obtain a larger (or smaller) answer than the computer offers. The program allows the teacher to assess the competence of the child in place value, and in the selected operation.

Another game, *BLOCKS*, is played by two children. Three random numbers indicated by dice are displayed, which are then used by the children to obtain a number available on a matrix (like a bingo card). The number obtained is highlighted in colour, and the children compete to place four numbers in a row on the matrix. At first, the numbers are quite easy to obtain. The three initial random numbers can be combined by any operation, but as the game proceeds the children need to aim at particular numbers left on the matrix, so they have to be flexible in their manipulation.

In *TRAINS*, designed for younger children, the train may have a number of coaches which seat a selected number of people in each. One task could be to distribute 21 people between 5 coaches seating 3 or 5 people, giving the solution: 3 coaches of 5 and 2 coaches of 3. The numbers are presented randomly and the teacher can select the level of difficulty. Again, a flexible approach to numbers is needed.

Skills

While it is not desirable to use programs which merely provide drill and practice, programs are available which allow skills to be practised while not being too mechanical. *GUSINTER* is a program which has children aiming to place numbers, four in a row. Only one number is presented on the screen at a time, and the children have to choose a multiple of that number. This 'game' is fun for the children and yet also provides good practice in recognizing multiples. It does not stop at multiples of 12; for example, it covers 51 as a multiple of 3.

On Microsmile, *TOWERS* offers a good approach to fractions. Each tower has to be larger than the one before but still less than 1. Because of the change of starting point the game becomes harder than one might think!

IDENTIFY and *DEFINE* are a pair of programs requiring the pupil to examine numbers and investigate their properties.

Investigations

Because the computer will draw and redraw diagrams so readily, it is an ideal tool for investigations. It lends itself to the idea of trying small numbers and then progressing to larger numbers. The computer will also make tables of results. This process helps to keep an investigation going and motivates the pupils.

Examples to start with are *THE MYSTIC ROSE* from Microsmile and *QUILTS*. Also recommended are the Anita Straker Investigations such as *SQUARES* and *DIAGONALS*.

Pupils should be encouraged to do some practical work in parallel with the work on the computer, so that the two aspects of the investigation are combined.

Art

Most children, even if they are not good at art, enjoy drawing on a computer. The special effects (spray, fill, etc.) are more fun than painting on paper. Most art packages have facilities for manipulating shapes, changing the size and orientation. They include facilities for using flip (reflection), rotation, transformation, tessellation, symmetry, etc., e.g. *IMAGINE*. The teacher can therefore use this enjoyable exercise to draw attention to the geometric concepts in the mathematics curriculum.

Concepts

Some computer programs concentrate on only one aspect of mathematics. If presented as a game, the competitive aspect, competing against the computer, encourages the pupil to persevere.

The Microsmile program *FACTORS* demands a strategy of selecting numbers by considering factors. Using this program, three or four tries make a pupil aware of the factors of the numbers selected. Another Microsmile program concentrates on reflection: the pupil needs to draw reflections across different mirror lines, and has to be constantly aware of the direction of movement required.

Special needs

Nowadays, pupils with special needs tend to be integrated into ordinary classes and, to meet their requirements, the teacher has to provide individual programmes of work for them. A computer can be a very useful resource in these circumstances because many programs are written specifically for pupils with special needs. In addition, many programs of general use can be set at different levels – sometimes allowing these children to use the same program as others in their class.

Some programs use graphics in such a way that helps pupils with special needs to understand the manipulation of numbers or shapes. There are also programs which enable a problem to be broken down into smaller units – for example, *TEN TRUCK* and *CAR PARK*.

More difficult programs

There are some substantial mathematical packages (*NUMBER GRIDS*, *LOCUS*, *NUMERATOR*) which are more for use in secondary schools but may be suitable for high attainers (Levels 5 and 6) in primary schools as well.

High-attaining pupils also have special needs and require appropriate provision for work at a higher level than their classmates. They benefit from individualized work, such as can be made available through programs like *NUMERATOR*.

Selection of programs

When selecting programs the teacher needs to be aware of her aims in using them. She should select a group of programs to use in class with which she feels comfortable and which can be used over the range of abilities of the children. The choice of programs will be limited by the hardware available.

The number and variety of programs are increasing, especially for use with the newer computers. Better graphics are possible on these machines, which have larger memories, higher speed of access and monitors with higher resolution. The programs selected should include those with a flexible open-ended quality, such as investigations and adventure games, that ensure the children have to think mathematically.

Note: The programs given as examples are well-established ones, which should be familiar to the reader. It is recognized that there are now more programs available, allowing a greater choice.

Chapter 9

Using LOGO for Key Stages 1 to 3 of the National Curriculum

INTRODUCTION

With the introduction of the National Curriculum there has been more emphasis on classroom assessment. There are two ways of assessing children's progress:

- either teachers give them a series of tasks which they expect children to be able to perform, and note their successes and failures; or
- they give them one task which can be carried out at several different levels and note the level reached by each individual.

The teacher also has to be aware of progression through the key stages.

LOGO is a convenient medium for the assessment of many mathematical concepts, since the children's thinking can be analysed from their performance. If LOGO is used, it not only satisfies many Programmes of Study in the mathematics curriculum, but it also enhances the pupil's understanding of the basic mathematical concepts underpinning those statements.

The experience of teachers who use LOGO regularly with young children is that they use more mathematical strategies than their teachers expected. They gain confidence in using degrees for angles and in operating on them. They work with large numbers, add and subtract, halve and double them in their heads. They also use calculators and know which calculation they need to do. They learn right from left, not just when the turtle on the screen is vertical, but from any position of the turtle. They learn to estimate length and angles remarkably well.

Within the framework of the tasks described in this chapter, there is scope for individual work at different levels of sophistication. Teachers should be prepared to encourage the children to develop their ideas as far as possible. Thus the levels reached by the children will vary according to how they manage the tasks undertaken.

Learning LOGO

LOGO is a language for children. Very few classroom teachers were taught LOGO in their own schooldays, and therefore they will have to learn it themselves before they can use it with their pupils. It is an advantage if teachers learn LOGO in the same way as their pupils, experiencing some of the excitement of exploring its capabilities and becoming aware of its limitations.

Adults often apply the Euclidean geometry they were taught at school to the use of LOGO and start by drawing a square. Although many principles of Euclidean geometry can easily be demonstrated using LOGO, this is not the way to introduce turtle geometry, which is more akin to navigation. Instead, it is necessary to think in terms of movement and bearings; pupils drawing with LOGO need to consider how to walk along a path or around a shape. It develops their spatial awareness.

For this reason it is usually best to start on Levels 1 and 2 with some work in PE on commands and directions. Playing 'People LOGO', one person directs a partner to make certain moves (this tactic is equally valid for adults and children):

Take 4 paces forward.
Turn to your right.

and so on. The first LOGO skill to learn is that the linear commands (forward and backward) must be kept separate from turning commands (left and right). In particular, it is necessary to learn to specify the amount of turn needed for the person to be facing in the correct direction before making the next move.

This can be followed by work using a robot such as a Roamer or a Pip or a floor turtle. While 'People LOGO' can always be carried out (the school having no shortage of available people), robot activities depend on available resources. Using robots is a requirement of the Information Technology Curriculum, and it is important that, even if 'real' robots are not available, children should have the opportunity to use LOGO.

Logotron

Although the LOGO commands vary on different computers, and even in different versions, many commands are common to all the languages. In what follows, the Logotron Language, one of the most common, is used. Because of the variation, it is essential that the teacher try out activities before using them in the classroom. For example, the 'default' setting for Logotron is WRAP, but on the Nimbus it is WINDOW. For Activity 9.5 the WRAP mode is needed. So if you are not using Logotron, type WRAP before you begin.

- WRAP – when the turtle comes to the edge of the screen it reappears on the opposite edge and carries on;
- WINDOW – when the turtle reaches the edge of the screen it carries on out of sight;
- FENCE – the turtle will not go past the edge of the screen and if given such a command will refuse to carry it out.

I find WRAP mode the best to use with young children as the turtle does not disappear nor does it refuse a command.

Some more or less open-ended activities that can be tried are now suggested and then their relationship to the National Curriculum is investigated. These activities are designed to provide ideas of what can be done with (teachers or) children – they are not to be slavishly copied!

LOGO ACTIVITIES

With all LOGO activities, the tasks should be done in small groups of two or three; it is useful for the players to be able to discuss the moves, and for the teacher to listen to their discussion.

Activity 9.1 'People LOGO'. This activity starts by the teacher talking about giving instructions involving directions and distances. Working in pairs, one person directs the other, using only oral instructions, to a particular point in the room. Then they exchange roles. Next, each player works out a complete sequence of moves *before* giving the commands to their partner. This requires far more skill than might be expected. Another interesting development of this task is to blindfold the person being directed, although some care may be needed to avoid accidents.

Activity 9.2 Using a Roamer (or another robot), players devise a set of activities or games. These can include such exercises as:

- sending the robot to a particular person or place;
- knocking over a toy or a bottle;
- using the robot to put a ball into a 'goal';
- using the robot as a bus or as a postman to move across a map or a model of a village;
- using the robot to draw a shape or a pattern;
- making the Roamer pass through an obstacle course.

This work is great fun and when the players have tried some of these activities, they will suggest many more!

This type of task should be repeated at intervals to consolidate work done on the computer. The programming of the robots can become quite sophisticated, as functions such as REPEAT are made available.

Activity 9.3 This involves designing a trail, using LOGO on a computer. Place three small pieces of Blu-tack randomly on the monitor screen. The task is then to use the turtle either to join the points marked by the Blu-tack or to make a shape to enclose all three markers. The exact challenge can be chosen and may be varied to suit different people. Use the commands FORWARD, BACKWARD, RIGHT, LEFT, and CLEAR SCREEN (abbreviated to FD, BD, RT, LT and CS).

Activity 9.4 This is an extension of Activity 9.3 but much more imagination can be used on this task! Fasten to the monitor an OHP transparent sheet with a map drawn on it. The aim is to get from one point on the map to another, while keeping to the road.

When making such a map, first place the clean transparent sheet on the monitor to mark the size of the available screen, and to mark the home position of the turtle. Then draw onto it the map design; this makes it easier for the player to position the map correctly when it is required.

Activity 9.5 This activity is concerned with the positioning of shapes rather than the structure of the shapes themselves. Load into LOGO a set of shapes already defined as procedures (see later) and saved in the memory, e.g. *SQUARE*, *CIRCLE*, *HEXAGON*, *TRIANGLE* and *PENTAGON*, *STAR* and maybe some irregular shapes. These shapes are then used to design a 'string of beads' (a repeating sequence) or to make a pattern.

ASSESSING PROGRESS

One benefit of using LOGO is the number of different mathematical concepts that can be covered within the activities. However, this also gives problems when it comes to assessment.

- Will all the children have displayed the same facility with the tasks?
- Were the outcomes different?
- What did the children actually achieve?

Listening to what they are saying provides the best clue to their current level of understanding. Also, it is essential to remember that the statements in the National Curriculum need to be met by children more than once and in different ways before a teacher can be certain that the pupil truly understands a given concept.

The mathematics used in Activities 9.1 to 9.5 cover Programmes of Study in Using and Applying Mathematics, Number and Algebra, and Shape and Space.

Programme of Study for Using and Applying Mathematics

LOGO can be used as material for a practical task. By listening to groups of children the teacher can find out if they are predicting accurately what will happen. Observing them as they use the robot, listening to their discussion and asking appropriate questions should enable teachers to make assessments for Levels 1 and 2.

Programme of Study for Number

During the 'People LOGO' activity, the children may demonstrate their ability to use numbers in a practical context. Similarly, when using the Roamer, they may be estimating and using numbers up to 10. They will need to use addition and subtraction as they work out the correct route for the robot; they are comparing lengths and also angles.

If they are using a 'one-key' LOGO, then they will be estimating in numbers up to 10. With normal LOGO, they will be estimating in 100s, often treating 100 as if it were a unit but writing the number correctly when entering the commands.

It is sobering to listen to children using LOGO, and to notice the way in which they are able to handle operations such as addition and subtraction with these numbers, and to estimate the distances and angles they require to complete the activity. Assessments of Level 1 and of part of Level 2 should be easy to observe. Some children will probably have used the ideas of halving and doubling, but they are less likely to have used quarters. When using the robot, they may discover the need for standard units.

Programme of Study for Number and Algebra

In Activity 9.5, if the children made a sequence with the shapes, this may well satisfy Level 1. Level 2 is covered more satisfactorily in the next set of activities.

Programme of Study for Shape, Space and Measures

The children will have been using ordinary words to describe position and to give instructions to move along a line. They will have demonstrated their ability to compare the shapes and the sizes of 2D shapes and to recognize different types of movement. Assessment of achievement at Levels 1 and 2 can therefore be made (except for those involving 3D shapes).

MORE LOGO ACTIVITIES

Apart from the mathematics being acquired, when using LOGO we also need to look at the Technology document, as the skills that children are acquiring in using LOGO are those of managing a programming language. They are also becoming familiar with the use of the computer.

If LOGO is to be used in the classroom as a mathematical tool for the children, many more challenges will be needed, as well as time to play with each activity and to explore their developing skills. The next activities introduce more explicitly the concept of programming, via the REPEAT command, but first we need to recognize the limits of the language.

Activity 9.6 This is a valuable opportunity to play with large numbers and to use the words associated with them. The task is to provide an answer for the child who asks: 'What is the biggest number I can use?'

- How big is the screen?
- Find out the largest number the computer will accept.
- Use any angle and any large number, e.g. RT 23, FD 5000.

NB Ensure the WRAP mode is in operation for this activity.

Activity 9.7 Use LOGO to draw a closed shape. (This does not necessarily mean to draw a square. It means any shape that returns to the starting position.) The player should recognize that one full turn was needed, and perhaps that the distance moved obeys some rule.

Activity 9.8 This task introduces the REPEAT command, a satisfying command for children to use because it is so economical – a great deal can be achieved with a minimum of typing! Using the shape from Activity 9.6 – or another one (it does not need to be a closed shape) – repeat it using the REPEAT command, e.g.

REPEAT 12 [instructions for shape]

Also use the shapes used in Activity 9.5 to create more patterns.

Activity 9.9 Draw a regular shape, i.e. with all sides the same length, e.g. equilateral triangle, square, rhombus, regular pentagon, and so on. Can you use REPEAT command to program the same shape more efficiently?

Programming structures

Although children can explore LOGO and enjoy themselves, if they are going to progress they need to learn something about the structures of programming. It is surprising how easily they pick up these simple concepts.

Procedures

The next important stage in using LOGO to be learned is making a 'procedure'. This is often described as 'teaching the computer another word or command'. The children are asked to draw a shape and give it a name, for example:

TO *NAME*
< instructions for shape >
END

The computer then tells us the *NAME* is defined. (If they are using Logotron on a BBC, it is easy to use the 'copy' facility to transfer the instructions into the procedure. On other computers, the instructions need to be written down in order to copy the program into the procedure on the edit screen.) Now, if NAME is typed in (rather than all the instructions needed), the shape defined will appear on the screen straight away. The great advantage of defining procedures is that they save time and effort; and a class can collect a bank (or library) of useful shapes. These can be loaded into the computer for everyone to use, at the start of any LOGO session.

Nested procedures and REPEAT *loops*

Now we use one procedure within another (this is called 'nesting'). For instance, suppose we have a set of procedures to draw the letters of the alphabet, called INITIALA, INITIALB, etc. These can be combined using two or three procedures to create more procedures making a monogram of every child's initials.

```
TO MYNAME
INITIALJ
INITIALA
INITIALL
END
```

will display JAL.

Similarly, we can use the REPEAT facility inside another REPEAT. For instance, use the REPEAT command to build a regular polygon and then use the repeat command with an extra command to make a string of polygons or to revolve the shape. Interesting designs, e.g. snowflakes, can be created. The whole subject of fractals starts from these simple ideas.

Activity 9.10 Make a decorative framework for a picture to be used in a word-processed story, or to make a tessellated pattern. Define procedures and use nesting and REPEAT to make your program as efficient as possible.

MORE ON ASSESSMENT

From the Programme of Study for Number and Algebra, children's performance so far and their discussion, teachers should be able to assess whether individuals have reached Levels 3, 4 or 5.

The pupil's performance in this programme of study can be assessed from the problems tackled. The complexity of any calculations is dependent on the precise nature of the work undertaken and could be as high as Level 5.

Programme of Study for Shape, Space and Measures

Levels 1 to 4 need to be considered when assessing work with LOGO in these tasks.

Programme of Study for Handling Data

It is unlikely that the tasks will need to be assessed for handling data unless the RANDOM function has been used.

MORE ADVANCED LOGO

The main structure of LOGO graphics has been encountered, but there are many more skills which could be acquired. It is worth looking at Level 5 of the Mathematics National Curriculum and the Technology documents, and deciding which statements could be accessible through LOGO. Teachers must also remember that to confirm an assessment, it is better to have observed that skill in more than one context.

When pupils are using LOGO in the classroom, other skills are learnt as they are needed. For example, moving the turtle without its drawing a line, using colour, sending the turtle back to the centre or to a corner. It is important that the pupils acquire skills which make them independent when using LOGO. The thinking skills are important too: the need to keep to a logical sequence, and the use of procedures to simplify the programs and make them more efficient. The order in which these skills are introduced depends on the needs of the class and of individuals, and on the other mathematical topics they are studying.

More programming concepts

Variables

This is a good introduction to algebra, and children think it is magic.

Another skill which pupils should learn is that of using a variable. The variable is identified by using a colon, and instead of using a fixed number, a word or letter represents the value needed. In this procedure (named *SHAPE*), *SIZE* is used as a variable.

```
TO SHAPE :SIZE
REPEAT 5[FD :SIZE RT 72]
END
```

Now type *SHAPE 50* to produce a regular pentagon with sides of 50 units. *SHAPE 100* will produce a larger pentagon.

Activity 9.11 Extend the *SHAPE* procedure, first by trying other values for *SIZE* but then to make different shapes:

- Use a variable for number of sides (instead of 5);
- Use a variable for exterior angle (instead of 72) and see what effect this has.

Coordinate values

Points on the screen can be defined by using numbers, called coordinates. To control where shapes appear on the screen, it helps to use a procedure such as SETPOS [X Y] which will move the cursor to this position before a shape is drawn. Using a procedure for a shape you have already designed, e.g. STAR 50, you can then place the shape in various positions on the screen, e.g.

```
SETPOS [0 100] STAR 50
```

Try other numbers, including negative numbers. Investigate which numbers can be used and where the stars are placed. Co-ordinate values can also be used to draw a shape.

Saving/loading

The saving and loading of work so that it can be retained (usually on disc) and then be used on another occasion is an important advantage of using computers and is a skill which should be acquired.

Random numbers

The use of the random function is an interesting command to investigate. It can be used to replace a variable. For example, you could define stars of random sizes.

Colour

The commands for colour vary with the type of computer and the version of LOGO being used. On some machines the Mode needs to be set before the LOGO session starts and the memory may become restricted. It is a skill particular to the type of computer, so the teacher must investigate the colour facilities of the available computers, before using it with pupils.

Turtle shape

Some versions of LOGO, such as Research Machines LOGO 2, allow the turtle to change shape and, in fact, to take more than one shape. All these shapes are capable of independent movement at different speeds. These features are worth investigating and give pupils scope for writing interesting programs.

SUMMARY

The following offers a brief summary of the role of LOGO within the mathematics curriculum.

Programme of Study for Using and Applying Mathematics (a tool for investigation and prediction)

- Level 2 Allows meaningful discussion.
- Level 3 Investigations and problem solving.
- Level 4 Identify problems and obtain information to solve them.
- Level 5 Interpret information.
- Level 6 The pupils can define their own problem/investigation.

Programme of Study for Number and Algebra

- Levels 1–3 Use numbers in practical situations. Double, halve, add, subtract, work with angles as well as length. Work in 10s and 100s, and learn place value.
- Level 3 Estimate distances and sizes of angles; multiplication and division by 2 and 3.
- Level 4 Place value.
- Level 6 Calculations in precise terms are often necessary in order to define a particular shape or sequence.

- Level 1 Repeating patterns, e.g. staircases.
- Level 2 Number patterns.
  Variables.
- Level 3 Often carrying out complex mental calculations by a variety of strategies.
  Inverse relationships.
- Level 4 Make and use patterns; pupils used to talking about LOGO.
  Simple formulae constructed in words; coordinates.
- Level 5 Follow instructions to generate sequences.
- Level 6 An investigation in LOGO can arise from a number pattern and may involve equations.

Programme of Study for Shape and Space

- Level 1 Use of turtle or robot for movement; comparison and ordering of shapes.
- Level 2 Description of shapes and types of movement.
- Level 3 Symmetry, description of shape.
  Points of the compass.
- Level 4 Construction of shapes.
  Estimation of distance and size of angles.
  Rotational symmetry.
  Work on area.
- Level 5 Use scale and measurement in units.
  Learn properties of shapes and the use of coordinates.
- Level 6 LOGO helps in the understanding of bearings and direction.

Programme of Study for Handling Data

Handling Data is not generally applicable at the lower levels.

- Level 4 Use of random numbers in LOGO to generate different patterns.
- Level 6 The use of the random function may help with the understanding of probability.

Answers

Note: In some of the examples the answers are obvious from the development of the problem. These answers have therefore not been included in this list.

Page 14 Activity 2.7
The goods cost £5.60; change is £4.40
The distances are 240 m, 1440 m, 560 m

Page 15 Ten times 13 is 130.

Page 23
1. 8 is worth 80 in 982; 8 is worth 8000 in 8045; 8 is worth 800 in 1807
2. For digits 0, 2, 9, the numbers in size order, smallest to largest: 029, 092, 209, 290, 902, 920.

Page 24
Patterns in addition square: 0 to 10
Major diagonal 0 2 4 6 8 10 12 14 16 18 20. Doubles of numbers 0 to 10
Minor diagonal 10. Pairs of numbers with sum 10
In rows and columns adjacent numbers increase by 1.
Analyse the number of times each total occurs in the addition table (1 to 6). The number of times is shown in brackets: 2, 12 (1), 3, 11 (2), 4, 10 (3), 5, 9 (4), 6, 8 (5), 7 (6)

Page 41 Activity 2.24
1. (i) 9.31 and 9 310 000
   (ii) 8140 and 0.0143
   (iii) 5 060 000 and 0.00506
   (iv) 3000 and 0.300
2. 5 800 000 is 6 000 000 to 1 sig. fig.
3. 358 $\div$ 29 is approximately 360 $\div$ 30 $=$ 12

Page 42
1. £4.50; £6.75; £11.25
2. 54 m; 47 cm; 75 ml; 135 g

Page 44 Activity 2.26
$x^3, x^4, x^6, x^6, x, x^3, x^2, x^4$

Page 45 Activity 2.27
1. 25%, 75%, $12\frac{1}{2}$%, $37\frac{1}{2}$%, $62\frac{1}{2}$%, $87\frac{1}{2}$%, $2\frac{1}{2}$%, 16.67%, 33.33%, 11.11%, 6.67%, 77.78%, 150%
2. $\frac{3}{4}, \frac{1}{8}, \frac{5}{8}, \frac{1}{3}, \frac{4}{25}, \frac{1}{4}, \frac{1}{40}, 1\frac{7}{20}, 1\frac{4}{5}, 2\frac{3}{20}$
3. £7, £10.50, £17.50
4. a) 90p; 60p b) £6; £4 c) £2.97; £1.98 d) £2.88; £1.92

Page 46

Fraction	Percentage	Decimal
33/50	66%	0.66
3/20	15%	0.15
7/20	35%	0.35
$\frac{3}{4}$	75%	0.75
$\frac{2}{3}$	$66\frac{2}{3}$%	0.67
1/7	14.28%	0.14
4/40	$7\frac{1}{2}$%	0.075
17/20	85%	0.85
1/8	$12\frac{1}{2}$%	0.125

Page 46 Activity 2.28

3) Mr Wong's VAT on £48 at 10% = £4.80; at 5% = £2.40; at 2½% = £1.20.
Total VAT = £8.40
Total bill = £56.40

5) Discount 10% of £12.50 = £1.25; final bill is £11.25
10% of £23.40 = £2.34; final bill is £21.16
10% of £37.50 = £3.75; final bill is £33.75

Page 82 Activity 4.3

a) 1256
b) 3600

Page 84 Activity 4.7

$n = 9$

Page 85

Activity 4.9

a) $1000k$ b) $60m$ c) $24d$ d) $100P$ e) $100M$ f) $1000K$ g) $M/60$ h) $G/1000$
j) $H/24$ k) $m/1\,000\,000$

Activity 4.10 10 km; $k = hs$

Activity 4.11 £ $N \div 6$

Activity 4.12 $P \div 4$

Activity 4.13 Perimeter = $2\ (2w + w) = 6w$; $f = 6w$; area = $w \times 2w = 2w^2$
(a) $f = 90$ metres; area = 450 m^2
(b) $f = 120$ m; area = 800 m^2

Page 86

Activity 4.14 (a) $C = 25.1$ cm; $A = 50.2$ cm^2 (b) $C = 45.2$ cm; $A = 163$ cm^2

Activity 4.15 25.5 cm; 254 cm; 5°C; 10°C; 20C°

Page 88 Activity 4.19

1. $c = 12n$
2. $P = 2(l+w)$ rectangle; $P = 4s$ (square);
3. Perimeter = $4s$ for a square, 48 cm. Perimeter = 2 (11 + 13) or 48 cm for the rectangle. Their perimeters are the same. They have different areas: square, 144 cm^2; rectangle, 143 cm^2

Page 120 Activity 5.20 Where values for the angles are not given, make up your own values, e.g. in (a) angle a could be 55°. State the value of b, giving your reasons for this. $b = 125°$ ($a + b = 180°$, because they are on a straight line.
(c) $d = 115°$; $g = 65°$, vertically opposite to c; $h = 115°$, vertically opposite to g
(d) $b = 135°$, adjacent to a
(e) $b = 45°$, alternate angle; $a = 70°$ (3rd angle on a straight line); $c = 65°$, alternate angle
(f) $p = 100°$ (3rd angle of triangle), $b = 50°$ (corresponding angle), $a = 100°$ (alternate angle with p)
(g) $c = 115°$, adjacent to 65°

Page 123 Activity 5.24
1. All three angles 60°
2. Angles 90°, 45°, 45°
3. 100°, 40°, 40°
4. 30°, 30°, 120°

Activity 5.25
$a + b = 90°$
$a + 2a = 90°$; $3a = 90°$; $a = 30°$, $b = 60°$, $c = 90°$

Activity 5.26
Regular pentagon 3 triangles
Total sum of angles = $3 \times 180° = 540°$
Each of 5 equal angles = $540° \div 5$ so each angle = 108°

Page 124 Activity 5.27
<BAC = a, <ABC = b, <BCA = c, <ACP = 180° (3 angles on a straight line)
$a + b + c = 180°$

Page 125 Activity 5.29
Dimensions of cuboids in cm, all of volume 36 cm
$1 \times 1 \times 36$ $1 \times 2 \times 18$ $1 \times 3 \times 12$ $1 \times 4 \times 9$ $1 \times 6 \times 6$

Areas of circles
i) 1963.75 cm^2
ii) 141.04 cm^2
iii) 13071.52 cm^2

Page 126
Volume of cylinders
i) 452.45 cm^3
ii) 132.75 cm^3
iii) 273.5 cm^3

A rhombus is a parallelogram with one pair of adjacent sides equal

Page 133 Activity 5.35
regular pentagon: interior angle 108°, exterior angle 72°
regular hexagon: interior angle 120°, exterior angle 60°
regular octagon: interior angle 135°, exterior angle 45°
regular nonagon: interior angle 140°, exterior angle 40°
regular decagon: interior angle 144°, exterior angle 36°

Page 151 Activity 6.4
Average speed 8 km per hour

Page 187 Activity 8.1
2, 4, 6, 8 . . .
1, 3, 5, 7 . . .
To reach 37 you need the sequence that generates odd numbers. To reach 100 you need the sequence that generates even numbers

Activity 8.3

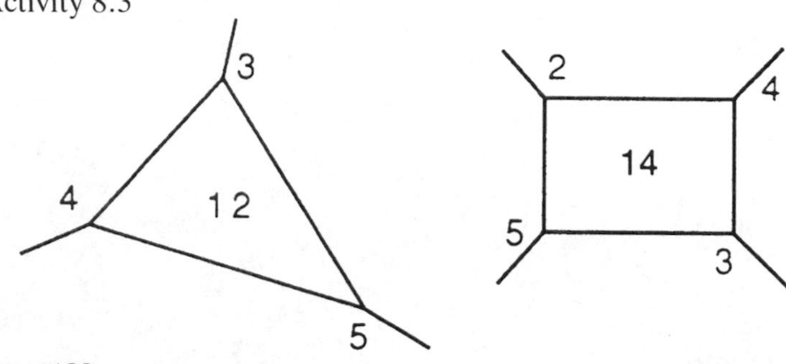

Page 188
Activity 8.4 20, 18, 16, 14 . . .
23, 21, 19, 17 . . .
25, 22, 19, 16 . . .

Page 189
Activity 8.6 $3 + 2 = 5$
$5 - 2 = 3$
$6 + 3 = 9$
$6 - 3 = 3$
$6 \times 3 = 18$
$6 \div 3 = 2$

Activity 8.8 0, 3, 6, 9, 12 . . .
12
12, 9, 6, 3, 0
3
0, 3, 6, 9, 12
12

Page 190 Activity 8.9 3.3333333 (also an E)
E indicates an error because the calculator cannot complete the calculation.

Activity 8.10 Answer must be more than 800 and less than 1000.

Activity 8.11 36, 396, 3996, 39 996, . . . 39 999 996
The next one will be 399 999 996.

Page 191 Activity 8.12 3.46

Activity 8.13 3.16, 4.47, 5.48, 6.32

Activity 8.14 4.586, 7.414